S250 Science in Context
Science: Level 2

The Open University

TOPIC 7   Nanotechnology

Prepared for the Course Team by Shelagh Ross, Sam Smidt and Jeff Thomas

Conclusion to the course

Prepared for the Course Team by Pat Murphy and Richard Holliman

This publication forms part of the Open University course S250 *Science in Context*. Details of this and other Open University courses can be obtained from the Student Registration and Enquiry Service, The Open University, PO Box 197, Milton Keynes, MK7 6BJ, United Kingdom: tel. +44 (0)870 333 4340, email general-enquiries@open.ac.uk

Alternatively, you may visit the Open University website at http://www.open.ac.uk where you can learn more about the wide range of courses and packs offered at all levels by The Open University.

To purchase a selection of Open University course materials visit http://www.ouw.co.uk, or contact Open University Worldwide, Michael Young Building, Walton Hall, Milton Keynes MK7 6AA, United Kingdom for a brochure. tel. +44 (0)1908 858785; fax +44 (0)1908 858787; email ouwenq@open.ac.uk

The Open University
Walton Hall, Milton Keynes
MK7 6AA

First published 2006. Second edition 2007.

Edited and designed by The Open University.

Typeset by The Open University.

Printed and bound in the United Kingdom by Halstan Printing Group, Amersham.

ISBN 978 0 7492 1892 8

2.1

The S250 Course Team

Andrew J. Ball (*Author, Topic 2*)

John Baxter (*Author, Topic 6*)

Steve Best (*Media Developer*)

Kate Bradshaw (*Multimedia Producer*)

Audrey Brown (*Associate Lecturer and Critical Reader*)

Mike Bullivant (*Course Manager*)

James Davies (*Media Project Manager*)

Steve Drury (*Author, Topic 3*)

Lydia Eaton (*Media Assistant*)

Chris Edwards (*Course Manager*)

Mike Gillman (*Author, Topic 4*)

Debbie Gingell (*Course Assistant*)

Sara Hack (*Media Developer*)

Sarah Hofton (*Media Developer*)

Martin Keeling (*Media Assistant*)

Richard Holliman (*Course Themes and Author, Topic 1*)

Jason Jarratt (*Media Developer*)

Simon P. Kelley (*Author, Topic 2*)

Nigel Mason (*Topic 7*)

Margaret McManus (*Media Assistant*)

Elaine McPherson (*Course Manager*)

Pat Murphy (*Course Team Chair and Author, Topic 1*)

Judith Pickering (*Media Project Manager*)

William Rawes (*Media Developer*)

Shelagh Ross (*Author, Topic 7*)

Bina Sharma (*Media Developer*)

Sam Smidt (*Author, Topic 7*)

Valda Stevens (*Learning Outcomes and Assessment*)

Margaret Swithenby (*Media Developer*)

Jeff Thomas (*Author, Topics 6 and 7*)

Pamela Wardell (*Media Developer*)

Kiki Warr (*Author, Topic 5*)

The Course Team would like to thank the following for their particular contributions: Benny Peiser (*Liverpool John Moores University; Author, Topic 2*), David Bard (*Associate Lecturer; Author, Topic 6*) and Barbara Brockbank (*Associate Lecturer; Author, Topic 6 and Critical Reader*).

Dr Jon Turney (*University College London and Imperial College London*) was External Assessor for the course. The External Assessors for individual topics were: Professor John Mann (*Queen's University, Belfast*); Professor John McArthur (*University College London*); Dr Richard Reece (*University of Manchester*); Dr Rosalind M. Ridley (*University of Cambridge*); Dr Duncan Steel (*Macquarie University, Australia*); Dr David Viner (*University of East Anglia*) and Professor Mark Welland FRS (*University of Cambridge*).

Contents

Introduction

Nanoscience and nanotechnology are concerned with matter on an extremely small scale: one nanometre (1 nm) is one millionth of a millimetre, or 10^{-9} metres. For the purposes of this Introduction, nanoscale materials may be defined simply as those in which a dimension in the range 1–100 nm plays a crucial role.

As the 20th century gave way to the 21st, nanotechnology began to be trumpeted as bringing in a new industrial revolution. Neal Lane, at that time Assistant for Science and Technology to President Clinton and a former Director of the US National Science Foundation, speaking at a Congressional hearing in 1998 said:

> If I were asked for an area of science and engineering that will most likely produce the breakthroughs of tomorrow, I would point to nanoscale science and engineering.

Although nanotechnology is heralded as a turn-of-the-century development, its seeds were sown many decades before the word entered the dictionary. In 1959, the physics Nobel laureate Richard Feynman (1918–1988) gave a now famous lecture, which he entitled 'There's plenty of room at the bottom'. In this talk, he pointed out that many biological systems are exceedingly small, but can move, manufacture substances, repair themselves and make replicas of themselves. He envisaged a time when scientists, too, would be able to make objects that could manoeuvre and make new objects on an atomic or molecular scale. Then we would no longer have to do chemistry as a bulk process. Feynman was in no doubt that when (not *if*, note!) we gained control over the precise arrangement of things on an atomic or molecular scale, we would find that substances could have a much greater range of properties. Feynman was well known as a practical joker, and during this lecture he assured his audience several times that this vision was not a spoof.

> I am not inventing anti-gravity, which is possible someday only if the laws [of physics] are not what we think. I am telling you what could be done if the laws are what we think; we are not doing it simply because we haven't yet gotten around to it.

Well, it took 40 years, but we have got around to it now.

The hopes for nanotechnology are that the greater range of properties that Feynman foresaw will deliver a huge range of new applications, for example in the IT sector, the manufacture of 'smart' materials, the development of alternative energy sources, the clean up of pollution, medical diagnosis and drug delivery, among many others. It is on the basis of this kind of potential that a report entitled *New Dimensions for Manufacturing: A UK Strategy for Nanotechnology*, submitted in 2002 to the Minister for Science and Innovation, stated that

> Few industries will escape the influence of nanotechnology… It will affect so many sectors that failure to respond to the challenge will threaten the future competitiveness of much of the economy… [It] is difficult to estimate the size of the potential market, but it will be very large. Forecasts range from tens of billions to trillions of dollars.
>
> (Taylor, 2002)

Some of these developments are already with us, others are confidently expected in the next 5 to 10 years, while others are still only fantasy. As Sir David King, Chief Scientific Adviser to the UK Government, remarked in his foreword to a 2003 report on the social and economic challenges of nanotechnology

> science and technology enthusiasts and science fiction writers – sometimes indistinguishable from each other – have picked up on this new theme.
>
> (King, 2003)

In analysing reports of the potential of nanotechnology, especially – but not exclusively – those in the news media, it is important to be aware of where the science stops and the fantasy begins. Nanotechnology *has* already delivered some useful applications, although those that have been commercially produced so far have mostly been the result of evolutionary developments from previous technologies, rather than examples of revolutionary progress. In the medium term, it is confidently expected that nanotechnology will enable the production of many new materials and devices. The potential of nanotechnology for medical diagnosis and treatment is seen as particularly promising, although it may take a decade or more to bring to fruition, partly because of concern over associated risks which can only be investigated through lengthy testing. However, it is important to distinguish realistic hopes for future developments, and reasonable concerns about their possible risks and impacts, from exaggerated claims that have no firm scientific basis. Hyperbolic expectations (i.e. 'hype') of technological futures, in which, for example, machines and biological systems become incorporated into one another to the extent that they become indistinguishable, often have as their flip side equally hyperbolic fears, such as the prospect of self-replicating nanorobots reducing the entire biosphere to 'grey goo'. By the end of your study of this topic, you should be better equipped to see where the boundaries between hope and hype might lie.

Promise and risk are familiar partners, with many previous technologies having had unanticipated negative consequences. The effects of nanotechnology should not be expected to be any more predictable. One of the first serious concerns to be raised related to the possible toxicity of nanoscale particles. Largely on this basis, the ETC group, a Canadian organisation with a history of influential opposition to GM crops, used the occasion of a World Summit on Sustainable Development in 2002 to call for a moratorium on the development of nanostructured materials; its report entitled *The Big Down*, published in January 2003, was given widespread media coverage.

The public perception of both benefits and risk is crucial to the acceptance of new developments, as you have seen with GM. Many observers consider that the GM experience had a considerable bearing on attitudes to nanotechnology. The early signs of public concern about nanotechnology were interesting, in that they represented a backlash against a new technology that had hardly even emerged. In Britain, the major players responded proactively to the situation. In June 2003, the UK Government commissioned the Royal Society and the Royal Academy of Engineering to carry out a review of the current and potential developments in nanotechnology, with particular emphasis on its environmental, health and safety, ethical and societal implications. The resulting report, entitled *Nanoscience and*

Nanotechnologies: opportunities and uncertainties, was published in 2004 and the official government response in February 2005. Some of the conclusions and recommendations from both documents are discussed in the chapters that follow. In 2003, the Better Regulation Taskforce also called for the development of a new regulatory framework for nanotechnology, and for early and informed dialogues between scientists and the public about its potential impacts. A number of scientists, on both sides of the Atlantic, commented that, as a result of the GM debate, the UK was in 2005 already on the way to developing mechanisms for such dialogues, whereas scientists in the USA were less well prepared to participate in this kind of discussion. In 2005, an interesting first step in public engagement was taken by the establishment of a UK citizen jury on nanotechnology.

The field of nanotechnology is already vast and is developing extremely rapidly in many different directions. Unlike some of the earlier topics in the course, this is not a story with a clear narrative thread. Nor, within the space available in this course, is it possible to provide a comprehensive overview. So what you will find in the chapters that follow is mainly a series of examples chosen to illustrate some of the properties of nanoscale structures, the tools used to investigate them, applications that might derive from them, and some of the benefits and challenges to society that may arise as a result. Chapter 1 sets the scene by examining some of the basic physical science that underlies work at the nanoscale. In Chapter 2, you will look at how nanoscale properties are being exploited to develop new materials – driven by the need for increasing miniaturisation in electronics and computing, and by demands for energy efficiency and environmental remediation. Chapter 3 considers the interaction of nanotechnology with living systems from several viewpoints: the nanoscale 'machines' within biological systems that Richard Feynman was alluding to in his lecture 'There's plenty of room at the bottom', artificial nanomachines that may make use of biological material, and medical developments that could exploit nanotechnology for diagnostics or drug delivery. In all three of these chapters, you will weigh possible benefits against possible risks and consider the balance between hope and 'hype'. Chapter 4 takes up these issues again, and also looks in more detail at the communication of ideas among scientists and at the intended democratisation of science through pressure groups and the citizen jury.

Chapter 1

Working at the nanoscale

As you have seen in the Introduction, the rapid growth of both media interest and serious investment in nanotechnology is due to perceptions of its potential in terms of applications. It is arguably for this reason that the 'technology' label has stuck to the field, despite the fact that most of the suggested technological applications are, at the time of writing (2006), more a concept than a reality. As things stand now, much of the work being done under the umbrella of nano*technology* is in fact nano*science*, carried out by scientists in a wide range of disciplines, including physics, chemistry, biology, biochemistry and areas associated with medicine. One of the aims of this chapter is to begin to unpick what it is that makes nanotechnology so interesting and exciting *as science*. During this preliminary exploration of science at the nanoscale, the definition of nanotechnology will also gradually be refined and extended.

1.1 Scaling down

In the Introduction to this topic, nanoscale materials were defined simply as those in which dimensions in the range 1–100 nm play a crucial role. In this section, you will begin to get a feel for this length scale and to appreciate why it is of particular relevance in so many areas of science.

1.1.1 Micro to nano

The smallest division on an ordinary ruler is 1 millimetre: 1 mm = 10^{-3} m. A micrometre is 1000 times smaller than a millimetre: 1 μm = 10^{-6} m. A nanometre is 1000 times smaller still: 1 nm = 10^{-9} m.

■ 1 mm is the thickness of about 10 sheets of paper. What roughly is the thickness of one sheet in (a) metres, (b) micrometres, (c) nanometres?

▨ The thickness of 10 sheets ~1 mm = 10^{-3} m

(a) the thickness of one sheet ~$\dfrac{10^{-3}}{10}$ m = 10^{-4} m

(b) 10^{-4} m = 100×10^{-6} m = 100 μm (i.e. 10^2 μm)

(c) 100×10^{-6} m = $100 \times 1000 \times 10^{-9}$ m = 100 000 nm (i.e. 10^5 nm)

These thicknesses are placed in context in Figure 1.1, which shows a length scale ranging from 1 mm down to 0.1 nm.

The discovery, in 1985, of one of the examples in Figure 1.1 – the C_{60} molecule ('buckyball') – was a significant episode in the nanotechnology story. This breakthrough is described in Box 1.1.

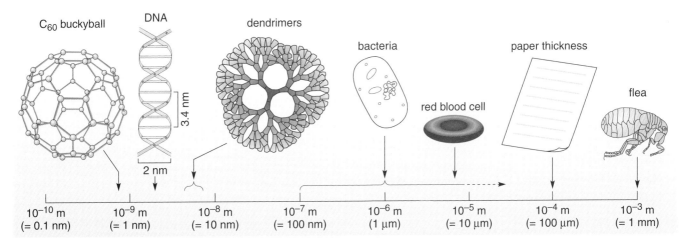

Figure 1.1   Length scale from 0.1 nm to 1 mm. Note that the scale is logarithmic. You will meet the C_{60} 'buckyball' shortly, and dendrimers later in this chapter.

Box 1.1 Discovery of a new form of carbon

A number of forms of carbon have been familiar for hundreds of years: the soft black particles of soot and charcoal, the graphite used in 'lead' pencils and the clear, hard gemstone, diamond. Carbon has four outer electrons; it forms covalent compounds (i.e. molecules in which the atoms share pairs of electrons). The carbon–carbon bond distance varies according to the number of electrons involved in the bonding. The more electrons involved, the tighter the bond and the shorter the bond distance. For a single bond (e.g. in ethane, H_3C-CH_3) it is 1.54 nm; for a double bond (e.g. in ethene, $H_2C=CH_2$) it is 1.34 nm.

Graphite (Figure 1.2a) consists of layers of regular hexagons of carbon. The bond distance in the hexagons (1.42 nm) is intermediate between those of single and double bonds, and the nature of the bond is also intermediate between single and double. The distance *between* the layers, however, is much larger, at 3.35 nm. The layers are held together not by strong covalent forces, but by much weaker London (van der Waals) forces. It is the weakness of the London forces that explains why graphite shears so easily parallel to the layers; this characteristic gives graphite its lubricating properties. This is an example of a structure with different properties in different directions. Such structures are called **anisotropic**. Diamond (Figure 1.2b), on the other hand, is **isotropic**: its properties are identical in each principal direction. Each carbon atom is bonded tetrahedrally, with all the bond lengths exactly as would be expected for single bonds (1.54 nm). This isotropic structure is

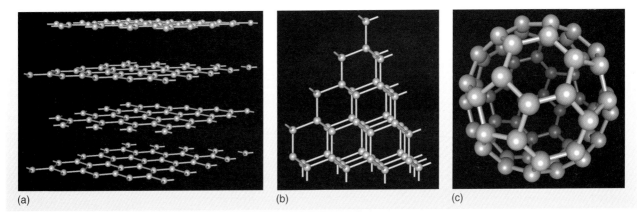

(a)   (b)   (c)

Figure 1.2   Three forms of carbon: (a) graphite, (b) diamond, (c) C_{60} molecule ('buckyball').

very strong: diamond is the hardest known naturally occurring substance, and is used in specialist cutting tools.

The structures of graphite and diamond have been familiar to many generations of chemists. But in 1985 a previously unknown form of carbon was discovered. Harry Kroto, then of the University of Sussex, and collaborators at Rice University in Houston, USA, tried to synthesise carbon clusters similar to those known from spectroscopic signals of molecules in the gas clouds of outer space. In the process they unexpectedly discovered a completely new form of carbon. Its molecules turned out to consist of 60 carbon atoms joined together to form a ball, in which 20 rings of six carbon atoms are connected to 12 rings of five carbon atoms (Figure 1.2c). C_{60} was named **buckminsterfullerene** – colloquially often shortened to 'buckyball' – after the American architect Buckminster Fuller (1895–1983), whose designs for geodesic domes had a similar geometry. A method for bulk synthesis of C_{60} soon followed. An electric arc is passed between two graphite electrodes in a chamber containing helium at low pressure. This causes intense local heating, which ruptures many C—C bonds in the graphite. Atoms and clusters of atoms evaporate from the surface and then cool in the helium atmosphere, resulting in the formation of large molecules. By 1990, sufficient quantities could be synthesised to confirm that the molecules were 75% C_{60} and 23% C_{70}, with a few molecules of higher mass (e.g. C_{120}). All these molecules contain 12 pentagons of atoms linking different numbers of hexagons (C_{60} has 20, C_{70} has 25), and are now collectively known as the **fullerenes**. Smaller fullerenes (down to C_{28}) can also be made.

For their discovery of the fullerenes, Harry Kroto and his American colleagues Robert Curl and Richard Smalley received the 1996 Nobel Prize for Chemistry.

Question 1.1 should help you to get a feel for the nanoscale and to connect it to everyday, visible experience.

Question 1.1

(a) Figure 1.3 shows three spheres of very different sizes. The average diameter of the Earth is 6371 km. The typical diameter of a tennis ball is 6.5 cm. The diameter of a C_{60} molecule is 0.7 nm. To the nearest order of magnitude, calculate the ratio of the diameter of the Earth to that of a tennis ball, then the ratio of the diameter of a tennis ball to that of a C_{60} molecule. (*Hint*: it is a good idea to start by converting all the diameters into the same units – metres are the most convenient.)

(a)

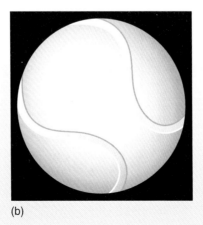
(b)

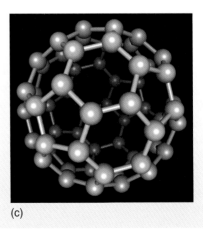
(c)

Figure 1.3   What are the ratios of the diameters of these objects? (Question 1.1a)

(b) The typical growth rate for a fingernail is 0.1 mm per day (the exact rate in an individual depends on a variety of factors, including age, sex, state of health and time of year). To one significant figure, estimate how much a fingernail grows in 1 second.

Both the prefixes 'micro' and 'nano' come from the Greek: *micros* means small, and *nanos* means dwarf. In the case of 'micro', the prefix is commonly used in an informal way simply to denote 'small', as well as in a scientific way to denote a factor of 10^{-6}.

■   List as many words as you can think of in a couple of minutes that have the prefix 'micro'. In which of these is the prefix clearly *not* used in this sense of scientific measurement?

▨   You may have thought of micron, micrometer, microscope, microbe, microfiche, microwave, microclimate, microcosm, microbiology, microsurgery, microelectronics, microprocessor, Microsoft™, microphone, microlight and probably a few others besides. The most obvious examples from this list in which the prefix is used in its informal sense of 'small' are microclimate (which is a term that may be applied to an area as small as a house wall or as large as a town), microcosm (literally 'small world') and microlight.

It is interesting to note that although certain *areas* of science are defined by falling into the microscale (microbiology for instance) there is no 'microscience' to encompass a range of investigations into phenomena in which dimensions on the scale of micrometres are crucial. So what is so extraordinary about phenomena on the scale of 1–100 nm that 'nanoscience' and 'nanotechnology' bring together many different specialities under one umbrella? One reason is that many branches of science, for example chemistry, biological and medical sciences, geoscience and physics, are now converging at the nanoscale.

Chemistry is the science that has traditionally been most concerned with the nanoscale. Chemists think about matter in terms of atoms and molecules, and the scale between 0.1 nm and 10 nm could be described as the typical atomic and molecular scale. Atoms cannot really be measured individually, because their electron clouds have no sharp edge, but an estimate can be made from the distance between the centres of two identical atoms in a molecule. On that basis, a hydrogen atom has a diameter of 0.06 nm. Fairly large molecules, such as C_{60} or sucrose ($C_6H_{12}O_6$), are about 0.7 nm across. Of course, chemists long ago synthesised molecules with one dimension much larger than 1 nm by joining together small molecular 'building blocks' (monomers) into long chains (polymers). The whole plastics industry is founded on the polymerisation process. Polymer chains typically have lengths in the tens of nanometres. However, other relatively recent developments have led chemists to explore different types of large and increasingly complex molecules. One of these developments was the discovery of the fullerenes (Box 1.1), which, as you will see in Section 1.4, was an important step on the way to making a special class of nanoscale materials. Another development has been the study of **supramolecular** chemistry, which is the science of complex assemblies of atoms and/or molecules held together by non-covalent interactions. Some of these assemblies are very large on a molecular scale, with dimensions in one or more directions of 10 –100 nm. There will also be more about supramolecules in Section 1.4.

While chemists are now working with larger entities than they did pre-1980, many biologists and medical researchers are increasingly concerned with subcellular structures and DNA molecules within the nanometre range. Proteins, for example, are between 1 nm and 20 nm in size; the separation between the strands in a DNA molecule is about 2 nm. Physicists, too, are very interested in the nanoscale, not least because at this scale materials can exhibit many novel physical properties, which you will find out about during your study of this topic. In addition, physicists have developed instruments for imaging surfaces so that individual atoms can be seen, as Section 1.2 will describe. Geoscientists are becoming very aware that nanoscale particles may play crucial roles in geochemistry, on Earth and on other planets; such particles are, for example, abundant on the Martian surface. One of the big hopes for nanotechnology is the synergy that could come from scientists in different fields sharing ideas and techniques so as to undertake interdisciplinary research. However, it is more easily said than done to find a common basis between scientists in very different fields. This is an issue that will be examined in more detail in Chapter 4.

A different kind of driver for research at the nanoscale comes from the increasing demands for miniaturisation in the electronics industry. The transistor, which is the basic component of the computer chip, was invented in 1948. In the fewer than six decades since then it has been reduced in linear size by nearly seven orders of magnitude, and hence by a staggering 14 orders of magnitude (i.e. $[10^7]^2$) in terms of area. As the working components become smaller, more and more components can be fitted into a given space on an integrated circuit. Gordon Moore, one of the founders of Intel, noticed a pattern to this trend in 1965 (even though at that time there were only 30 transistors on a silicon chip), leading him to propose an empirical rule that has now become known as **Moore's law**: the numbers of transistors on a computer chip doubles roughly every 2 years. Remarkably, Moore's law has continued to hold for the 40 years since 1965. The 'million transistor per chip' milestone arrived in the late 1980s with the Intel 486 processor. By 2005, the thousand million boundary was about to be breached. Eventually, however, a lower size limit will be reached at which devices produced by current miniaturisation techniques will cease to work because physical processes that are insignificant at larger scales will become important. For the continued development of the computer industry and all the other areas that depend upon information and communication technology, a pressing issue, therefore, is to understand the properties of matter on the nanoscale.

In Section 1.3 you will discover some of these properties, and in Chapter 2 you will see how they may be exploited in making useful new materials. First, however, Section 1.2 considers how we can visualise matter at such tiny scales.

1.2 Imaging the nanoscale

There are many devices for forming magnified images of objects that are far too small to be visible to the naked eye. This section discusses the physical principles underlying these devices and illustrates some of the work that can be done with them on nanoscale structures.

1.2.1 Limitations of optical microscopy

The simplest 'microscope' is the hand-held magnifying glass, which can be used, for example, to give an enlarged image of small print. More sophisticated versions, with an equivalent system of lenses, are used in the field by, for example, geologists and botanists: these hand lenses typically offer magnifications of 10 to 15 times. The compound microscope is a bigger instrument, with two lenses (or lens systems) mounted at opposite ends of a tube, one further magnifying the enlarged image formed by the other. The object under investigation is illuminated by shining light either onto or through it.

Because optical instruments require visible light, their performance is limited by diffraction effects (Box 1.2). Light emanating from a single point in the sample isn't focused to an equivalent single point in the image, but forms a bright diffraction disc surrounded by increasingly faint rings (Figure 1.5a). If two points on the sample are so close together that their diffraction discs overlap, then their images blur into one another and it becomes impossible to tell them apart (Figure 1.5b).

Box 1.2 Revision of diffraction and interference

Diffraction is the name given to the phenomenon in which a wave spreads out when it passes through an aperture of size comparable to its wavelength. The amount of diffraction increases as either the aperture is made smaller (Figure 1.4) or the wavelength is made larger.

The effect of the superposition of two or more waves is called interference. At the meeting point, the waves may reinforce one another or cancel one another out, either wholly or partially. In the case of light waves, reinforcement would result in a bright area, whereas total cancellation would result in a dark area.

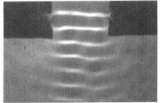 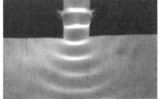 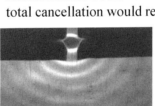

Figure 1.4   Diffraction of water waves through a gap in a barrier. All pictures are to the same scale; notice that the amount of diffraction increases as the size of the gap is reduced.

The **resolution** or resolving power of an instrument is the smallest separation between two objects at which the instrument can distinguish them as being distinct. This separation may be described in terms either of the angle the two objects subtend at the instrument or of the distance between the objects. The former is more usual in defining the resolving power of telescopes, and the latter in microscopy. Under the best possible conditions that can be achieved, when the distance between two points on the object is smaller than half the wavelength of the light being used then their diffraction discs in the image will overlap.

■   Given that visible light has wavelengths between about 400 nm (violet) and 700 nm (red), what in micrometres are the smallest structures that can be resolved with an optical microscope?

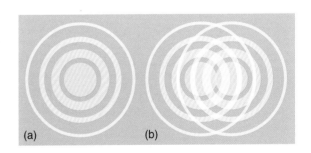

Figure 1.5   (a) Because of diffraction by lenses and apertures, the magnified image from a single point on the specimen is fuzzy, consisting of a bright ring surrounded by concentric faint rings. In this diagram the bright areas are coloured yellow. (b) When the image discs corresponding to two neighbouring points overlap, the points cannot be properly distinguished.

■ For visible light, half the wavelength is between 200 nm and 350 nm,
   i.e. 0.2 μm to 0.35 μm, and structures smaller than this would not be resolved.

Thus, optical microscopy can achieve a resolution of only a few tenths of a
micrometre at best. To see finer details than this it is necessary to use shorter
wavelengths. X-rays, for example, have wavelengths between roughly 10 nm and
0.01 nm, of similar size to the interatomic spacing in crystalline substances.
Diffraction of X-rays reflected from the regular planes of atoms within a crystal
can reveal their exact arrangement. However, much more direct ways to image
atoms on the surface of bulk materials are now available. Two of these methods,
scanning tunnelling microscopy and atomic force microscopy, are particularly valuable
in the nanotechnology context and are the subject of Sections 1.2.2 and 1.2.3.

1.2.2 Scanning tunnelling microscopy

Gerd Binnig and Heinrich Rohrer were awarded the 1986 Nobel Prize in Physics
for developing the **scanning tunnelling microscope (STM)**. This instrument is
based on the quantum mechanical phenomenon of tunnelling, by which an electron
can penetrate into regions that would be forbidden to it according to the laws of
classical physics. Although the details of the tunnelling process itself are beyond
the scope of this course, the principles of the STM are otherwise not too difficult
to understand. In this section, these principles will be described with schematic
diagrams. When you watch the movie that forms part of the activity at the end of
the chapter, you will see a real STM and get a better idea of its size.

Scanning tunnelling microscopy requires the substance under investigation to be
electrically conducting. A metal like copper is an obvious example of such a
conductor. Metals conduct electricity because they contain so-called 'free'
electrons. The electrons that are in inner shells remain close, and tightly bound, to
individual nuclei. However, electrons that are in outer shells are further away from
the nucleus and are also shielded from its electrostatic attraction by the inner
electrons, so they are much less tightly bound. These outer electrons are the
valence electrons, responsible for chemical bonding when the metal atoms
combine with atoms of other elements to form a compound. In a pure metal, the
valence electrons are freed from their parent atoms, forming an 'electron gas' that
pervades the whole metal; it is referred to as a gas because the valence electrons
can wander through the metal, much as the molecules of a gas move randomly
throughout the volume of a sealed container. The atomic nuclei with their
surrounding tightly bound electrons are thus left with a net positive charge, i.e.
they can be thought of as positive ions. These ions arrange themselves in a regular
array, called a **lattice**. The bonding in a metallic solid can thus be thought of as
arising from the attraction between this lattice of positive ions and the electron gas
occupying the same space. If one end of a metal wire is made positive with
respect to the other, then the free electrons will move in response to the voltage
difference, resulting in an electric current. The electron gas is bound only by the
surface of the metal itself: the electrons are confined behind an energy barrier
which prevents them escaping from the surface. According to classical physics,
which views electrons solely as particles, the electrons in the electron gas can
never leave the metal unless they are somehow given sufficient energy to
surmount the barrier. However, quantum mechanical understanding shows that
electrons can have wave-like properties. (You should recall that electromagnetic

radiation exhibits a similar wave–particle duality: in some situations, such as diffraction, light behaves like a wave, in others, such as the photoelectric effect, it behaves like a particle.) A direct consequence of these wave-like properties is that the least tightly bound electrons (in other words, the most energetic ones) can tunnel *through* the barrier presented by the metal surface and, therefore, are not necessarily confined within the metal just because they cannot go *over* the barrier. This 'escape' phenomenon is known as **quantum mechanical tunnelling**. A full understanding of tunnelling requires mathematics beyond the scope of this course, but simply being aware that it happens is all that is necessary in order to follow the basics of scanning tunnelling microscopy.

In an STM, a metal probe with an extremely fine tip is positively charged and positioned just above the electrically conducting surface to be imaged. The gap between the probe tip and the surface has to be very carefully adjusted: it must be large enough that normal electrical conduction cannot take place, but also small enough for electrons at the surface of the specimen to tunnel through to the tip. In practice, this means that the tip is only a few tenths of a nanometre from the surface. The flow of tunnelling electrons is measured as an electric current, called the 'tunnelling current'. Interestingly, this current is also nanosized, typically being a few tenths of a nanoampere. The probability of tunnelling depends both on the atom involved and on the separation of tip and surface: the probability falls off very rapidly with increasing tip-to-specimen distance. As a result, changes of only 0.01 nm in separation cause measurable changes in tunnelling current. If the probe is moved horizontally across the surface (a process referred to as scanning), then the tunnelling current will vary continuously, as illustrated in Figure 1.6. On this kind of scale, the surface of even an atomically smooth specimen (i.e. one in which the surface atoms are all in the same plane) of metal is contoured in hills and valleys. When the tip is directly above the 'bump' of an atom, the current will be higher than when it is positioned above the 'hollow' between atoms. By scanning in several directions, therefore, it is possible to build up a 'map' showing the arrangement of atoms on the surface of the specimen, with a resolution of better than 0.1 nm.

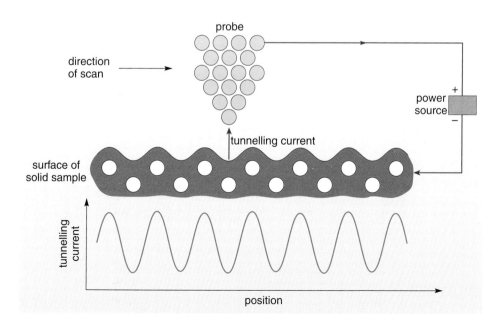

Figure 1.6   Variation in tunnelling current as a function of probe position along a horizontal line. Atoms are schematically represented as 'balls'. The current is at a maximum when the probe tip is closest to an atom on the surface of the specimen.

In practice, STMs are usually operated in constant-current mode. The tip is scanned across the surface and the distance between the tip and the surface is continually adjusted by a computer to maintain a constant value of the tunnelling current. If the current begins to increase, then the tip is automatically raised; if the current begins to decrease, then the tip is lowered. This is no mean feat given the tiny size of the tunnelling current! Control is achieved by mounting the probe on three tiny posts mutually at right angles and made of a special ceramic material, which expands or contracts on a scale of tenths of nanometres when an appropriate voltage is applied across it. (Materials that respond to a voltage across them by changing their shape, or conversely respond to deformation by generating a voltage across themselves, are termed **piezoelectric**.) The arrangement is illustrated schematically in Figure 1.7.

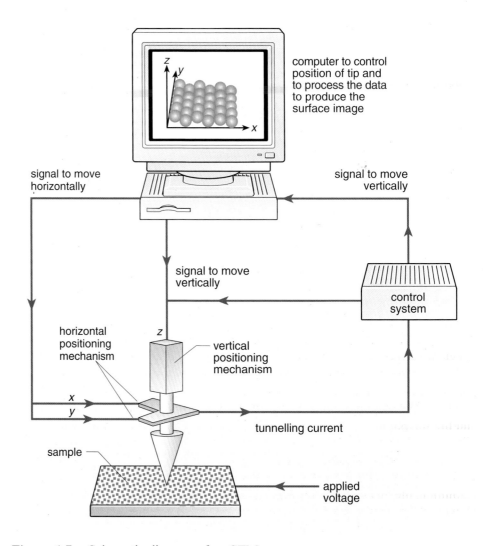

Figure 1.7   Schematic diagram of an STM.

Ideally, the system maintains the current at a constant value to within a few per cent, in which case the distance between the tip and specimen varies only by a few thousandths of a nanometre. The record of heights at different points across the surface is then represented as a topographic map that shows where the atoms protrude from, and where there are dips in, the surface. Figure 1.8 is an example of such an image.

Figure 1.8 An image formed by an STM of a nickel surface covered with a layer of sulfur atoms. Individual atoms and undulations on the surface on a 0.1 nm scale can be easily seen.

There are three main problems to overcome in an STM. The first is that it is very difficult to prepare surfaces that are absolutely smooth on the atomic scale. Surface undulations are clear in Figure 1.8: the pattern of light and dark is not uniform across the image. Since the tip follows *any* undulations in the surface, it can be difficult to disentangle effects of surface composition from those of the surface profile. Also, since the tip is only a few tenths of a nanometre from the surface, great care has to be taken to avoid it hitting the specimen, especially if the surface is not atomically smooth. That is one of the reasons for using constant-current operation, rather than constant height. In constant-current mode, irregular surfaces can be mapped very precisely, but the data take longer to build up because the system has to move the probe up and down. Scanning in constant-height mode is a faster operation, but is only feasible for very smooth surfaces. The second problem is that the very small separation between the tip and the surface means that the whole experiment is extremely sensitive to vibrations. Again, this problem is reduced by operating in a constant-current mode. The third challenge is the preparation of the probe tip. Tips are typically made of tungsten, platinum or a platinum–iridium alloy. To obtain the maximum resolution, the tip must be extremely sharp, ideally ending in a single atom! However, the smaller the tip, the more it vibrates; so, a balance must be found between the tip diameter and operational practicality.

STMs have the potential to be used not just to *image* nanoscale structures, but also to *make* them, while simultaneously picturing the new product. Figure 1.9 shows various stages in positioning a ring of iron atoms onto a copper surface with an STM tip. In this case there is no chemical bonding between the iron and

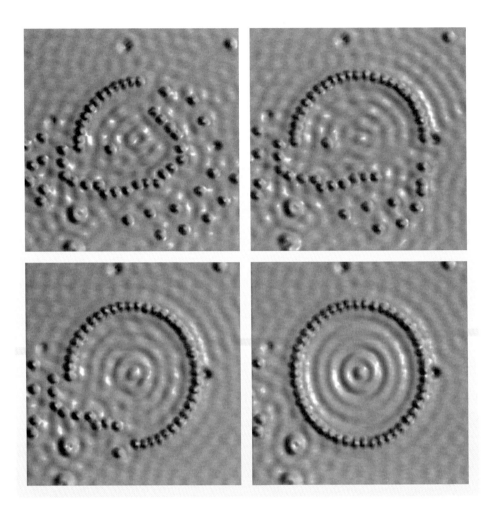

Figure 1.9   Making a 'quantum corral': the pictures show 48 iron atoms gradually being assembled by an STM tip to form a ring 14.3 nm in diameter on a copper surface (bottom right). The circular ripples within the completed ring show the wave nature of the electrons at the surface of the copper. Electrons are confined by the barrier of the iron atoms, resulting in a quantum-mechanical interference pattern as the electron waves bounce off this barrier, much as water waves would form an interference pattern if they were reflected by a solid vertical surface at the edge of a circular pool.

the copper – the iron atoms naturally rest in the hollows between adjacent copper atoms. It is important to realise that the iron atoms are not actually picked up by the tip; that would compromise the imaging function of the instrument. The tip is simply used to prod or pull the atoms gently from one hollow to the next without touching them, until they are in the required position.

Techniques for manipulating individual atoms or molecules in this way are developing fast. **Positional assembly**, literally building structures by putting individual atoms into the right places, is a key to one of the radical suggestions for the future of nanotechnology – molecular manufacturing. Traditional manufacturing methods can make smaller and smaller structures by machining or etching. This is referred to as a **top-down approach**. However, there is a fundamental limit to the miniaturisation that can be achieved by such techniques. Molecular manufacturing takes the opposite route, a **bottom-up approach** that aims to build both organic and inorganic structures atom by atom, molecule by molecule. Chapter 4 highlights a passionate debate about the feasibility of molecular manufacturing.

In the second half of the 1990s, considerable progress was made in handling and positioning objects on a nanometre scale and at room temperature. One project that received a great deal of media attention was the construction – by James Gimzewski and colleagues at IBM's Zurich Research Laboratory – of the 'molecular abacus' illustrated in Figure 1.10. The 'beads' of the abacus are C_{60} molecules, and they were positioned by an STM tip, along steps a single atom wide on a copper surface. These steps keep the molecules in line, in a way reminiscent of the ancient Roman form of the abacus in which the beads slid along grooves, rather than on wires as they do in the more familiar Chinese versions. When you watch the movie that forms part of the activity at the end of this chapter, you will see James Gimzewski talking about the nanoabacus.

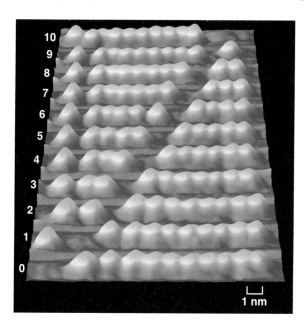

Figure 1.10 The nanoabacus, in which C_{60} molecules have been arranged along steps one atom wide on a copper surface. The lines of molecules count from 0 to 10.

1.2.3 Atomic force microscopy

The **atomic force microscope (AFM)** is also a device in which a sharp tip is scanned over the surface of a sample, but instead of measuring a current between the tip and the specimen it detects changes in the attractive or repulsive force between them. The heart of the instrument is a cantilever, on which the tip is mounted, as shown in Figure 1.11a. As the force on the tip changes, so does the vertical deflection of the cantilever, in the same way that a springboard would be depressed to different extents by divers of different weights standing on its free end. The magnitude of the force between the tip and the sample is very small, of the order of 10^{-11} N; so, although the cantilever is tiny (typically no more than 100 µm long), the deflection is also very small and difficult to measure. In the first AFM, which was made in 1985, the tip was a shard of diamond and it was attached to a strip of gold foil that acted as the cantilever. In this early device, the vertical deflection of the cantilever was actually detected by using an

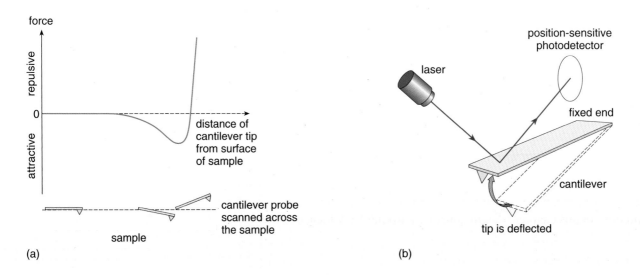

Figure 1.11 Principle of the AFM. (a) Schematic diagram of cantilever movement. (b) 'Beam bounce' method of measuring the vertical deflection of the cantilever.

STM above it. Nowadays, AFM tips and cantilevers are usually microfabricated from silicon or a silicon compound and the deflection is detected optically, as illustrated in Figure 1.11b. A laser beam is deflected off the cantilever into a position-sensitive photodetector which indicates the angular deflection of the cantilever. Data on the variation in force as the tip is scanned across the sample can then be converted into an image of the surface.

The AFM has a number of advantages over the STM. One is that, because what is being measured is a mechanical force and not a current, an AFM can give images of non-conducting surfaces. It can also work in a liquid as well as in air, and on a wide variety of samples, hard or soft.

Aside from their original use in AFMs, nanoscale cantilevers are now being used without tips as sensors for many kinds of chemical reaction. One ground-breaking experiment reported in the year 2000 involved coating the surface of a cantilever with short pieces (12 bases) of single-strand DNA. Other DNA strands with matching base pairs will bind to the original strands, exerting a force that bends the cantilever. The more pairs that match up, the greater the deflection. The technique has proved sufficiently sensitive to detect the presence of just one altered base within a short single strand of DNA. This technique is therefore seen as having great potential as a diagnostic tool for genetic disease; even the type of single base changes mentioned in Section 2.2 of Topic 6 could prove readily detectable.

1.3 Size matters: the properties of nanoparticles

Materials that are structured on the nanoscale can have novel properties, different from those of the same substance in bulk form. This arises because every physical property of a material has a characteristic length scale associated with it. One example of a characteristic length scale relates to electrical conduction. You have seen in Section 1.2.2 how electrical current in a metal wire corresponds to the movement of the valence electrons in response to a voltage difference across the ends of the wire. There is, however, some resistance to this movement, caused by the electrons being scattered away from the direction of flow when they collide with ions or impurities in the lattice. The characteristic length associated with this process is the mean distance travelled by the electrons between scattering events. In copper at room temperature, this length is of the order of 45 nm. Many characteristic lengths for physical processes are in the nanometre range, and new properties may arise if particles have dimensions that are similar to or smaller than these characteristic lengths. Properties are also strongly influenced by the number of dimensions that are of nanometre scale. For example, in Chapter 2 you will encounter 'nanowires' – structures that have a diameter of perhaps 40 nm, but can be grown with lengths of up to a millimetre. In this chapter we will consider particles for which all dimensions are of nanometre scale.

A particle measuring between 1 and 100 nm in all directions is generally considered to be a '**nanoparticle**'. However, a definition based purely on size is not entirely satisfactory, because many single molecules, especially biological ones, would fall within it, and not all large molecules are true nanoparticles. What distinguishes a nanoparticle is that its small size endows it with some properties that are not observed in the same material when it is in bulk form. In this section you will see why it is that nanoparticles have unusual properties and how those properties can be exploited in certain types of product. You will also discover why concerns have arisen over possible effects on human health from the manufacture of nanoparticles.

1.3.1 Surface effects

For many applications, the ratio of the surface area of a material to its volume is a very important property. Note that this is always referred to by physicists and chemists as the 'surface-to-volume ratio', with 'area' being implicit.

The surface area of a sphere of radius r is $4\pi r^2$ and its volume is $\frac{4}{3}\pi r^3$.

■ How does the surface-to-volume ratio of a sphere depend on its radius?

▨ $$\frac{\text{surface area}}{\text{volume}} = \frac{4\pi r^2}{\frac{4}{3}\pi r^3} = \frac{3}{r}$$

(Remember that, to divide by a fraction, you multiply by its inverse; so

$$\frac{4}{\frac{4}{3}} = 4 \times \frac{3}{4} = 3 .)$$

Clearly, the smaller r, the larger the surface-to-volume ratio. The next question is how the surface-to-volume ratio changes if a given volume of material is divided up into a number of smaller pieces.

■ Imagine a sphere of radius $R = 1$ cm. Approximately how many tiny spheres of radius $r = 1$ nm could be made from this volume of material?

▨ Suppose there are N tiny spheres. The total volume is constant; so

$$\frac{4}{3}\pi R^3 = N\frac{4}{3}\pi r^3$$

i.e. $R^3 = Nr^3$

So $N = \dfrac{R^3}{r^3} = \dfrac{(1\,\text{cm})^3}{(1\,\text{nm})^3} = \dfrac{(10^{-2}\,\text{m})^3}{(10^{-9}\,\text{m})^3} = \dfrac{10^{-6}\,\text{m}^3}{10^{-27}\,\text{m}^3} = 10^{21}$

■ If all the tiny spheres ($r = 1$ nm) were spread out so that their entire surface was accessible, how many times bigger would their total surface area be, compared with that of the larger ($R = 1$ cm) sphere? (*Hint:* remember to work in a consistent set of units.)

▨ The surface area of the big sphere, S, is

$S = 4\pi R^2 = 4\pi(1\,\text{cm})^2 = 4\pi(10^{-2}\,\text{m})^2 = 4\pi \times 10^{-4}\,\text{m}^2$

and that of a tiny sphere, s, is

$s = 4\pi r^2 = 4\pi(1\,\text{nm})^2 = 4\pi(10^{-9}\,\text{m})^2 = 4\pi \times 10^{-18}\,\text{m}^2$

There are 10^{21} tiny spheres, so their total surface area, s_t, is

$s_t = 10^{21} \times 4\pi \times 10^{-18}\,\text{m}^2 = 4\pi \times 10^3\,\text{m}^2$

So $\dfrac{s_t}{S} = \dfrac{4\pi \times 10^3\,\text{m}^2}{4\pi \times 10^{-4}\,\text{m}^2} = 1 \times 10^7$

The surface area has increased by a factor of 10 million!

(*Note:* here, all lengths have been converted into metres, but you could have chosen to work in centimetres or nanometres. Any *consistent* set of units is equally acceptable for this kind of calculation. You should, of course, still obtain the correct ratio at the end!)

Catalysis, especially heterogeneous catalysis, is one field in which the surface-to-volume ratio is of particular importance (Box 1.3). The greater the surface area of a heterogeneous catalyst, the more efficiently it works. For this reason, such catalysts are usually in the form of finely ground powders; but a catalyst made of nanoparticles should speed up a reaction even more than the same volume of the same catalyst in conventional powder form. The following activity will allow you to examine in quantitative detail the surface area that could be made available on the iron catalyst discussed in Box 1.3.

Box 1.3 Revision of Le Chatelier's principle and catalysis

Many chemical reactions require an input of energy if they are to proceed at a reasonable rate. For example, the reaction of nitrogen and hydrogen to form ammonia (an essential step in the production of artificial fertilisers) presents a dilemma to industrial chemists. The reaction

$$N_2(g) + 3H_2(g) \rightleftharpoons 2NH_3(g)$$

is reversible. This means that there is a state of chemical equilibrium (i.e. dynamic balance) between the forward reaction, which produces ammonia, and the reverse reaction, which breaks down ammonia. The relative proportions of reactants and product remain fixed so long as the conditions (e.g. the temperature and pressure) remain fixed.

Le Chatelier's principle is the basis for predictions about how the equilibrium will be affected by changes in the reaction conditions or constraints. This principle states that:

When a system in chemical equilibrium is subject to an external constraint, the system responds in such a way as to oppose the effect of the constraint.

The reaction of nitrogen and hydrogen to form ammonia is exothermic (i.e. it produces heat); so, by Le Chatelier's principle, it is favoured at lower temperatures. A high temperature favours the reverse reaction, whereby ammonia dissociates to form nitrogen and hydrogen. However, reducing the temperatures also results in the reaction rate slowing down. Even under the best compromise conditions,

the reaction proceeds slowly and with low yields of ammonia. Catalysis provides a way of helping things along. A **catalyst** does not affect the position of the equilibrium, but it does allow equilibrium to be reached more quickly. In this case, the use of a catalyst means that the reaction proceeds more quickly, so the temperature can be reduced, favouring the forward reaction at the expense of the reverse reaction and increasing the yield of ammonia. In the industrial production of ammonia, the catalyst used is iron. Because the reactants are in one phase (gas) and the catalyst is in another (solid), this type of catalysis is called **heterogeneous**. Since the catalyst is solid, all the catalysis takes place at its surface; thus, in order for the catalyst to be as effective as possible, it is prepared in powder form so as to have as large a surface area as possible.

Activity 1.1

Allow 45 minutes

A characteristic of nanoscale materials is that a greater proportion of the atoms are on the surface than is the case for bulk materials. This activity allows you to investigate this property in more detail.

In metals, there are three common ways for the atoms to be arranged to form a regular lattice structure. These arrangements are described in terms of a **unit cell**, which is the smallest repeating three-dimensional structure within the lattice. Figure 1.12a shows the unit cell for the so-called *body-centred cubic* (BCC) lattice: the unit cell is a cube containing an atom at its centre plus 'parts' of atoms that are shared with adjacent unit cells. In terms of space filling, the atoms in a BCC arrangement take up 68% of the volume of each unit cell. Iron is one of the metals with a BCC structure. In this activity you will work out the numbers of surface and interior atoms for various sizes of iron nanocrystals.

Because of the 'shared' atoms at the corners of a unit cell, Figure 1.12a is quite a difficult representation to work with. So for this activity you will instead build up a BCC lattice starting from a hypothetical cubic 'block' that has an atom at each corner and one in the centre of the block, as shown in Figure 1.12b. This representation shows the atoms as disproportionately small, but is simple to draw and to visualise.

The easiest way to begin is to think of various sizes of cubes made up of small numbers of these 'blocks' and simply count up the atoms in various positions. The thing to remember is that you are building a BCC lattice, so when blocks are assembled there is only one atom at any corner where the blocks touch.

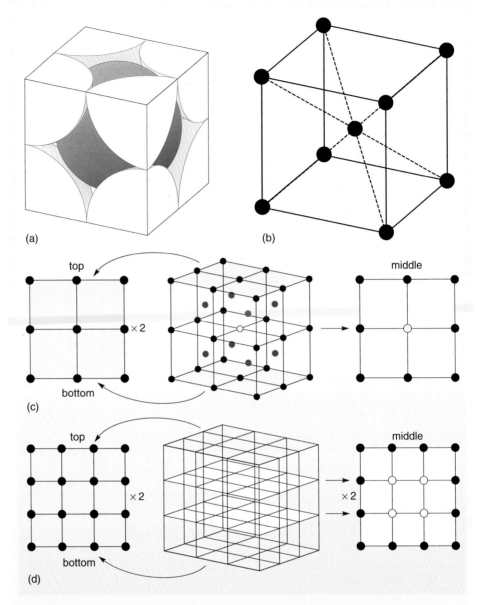

Figure 1.12   (a) The unit cell of a BCC metal; (b) hypothetical BCC-type block; (c) and (d) for use with Activity 1.1.

When there is just a single block, as in Figure 1.12b, there are eight surface atoms (one at each corner of the block) and one interior atom (at the centre of the block).

When a cube is constructed that is two blocks in each direction, there are eight blocks in all, as shown in Figure 1.12c. Now the atoms are most easily counted in layers. On each of the top and bottom layers there are nine atoms (3×3, coloured black), all of which are surface atoms. The middle layer (yellow) has eight surface atoms (coloured black) and one interior atom (coloured white). And each of the eight blocks has its central atom (coloured red), all of which are interior atoms. So the total number of atoms for a cube with eight blocks ($2 \times 2 \times 2$) is

	Surface	**Interior**
Top and bottom layers	2×9	0
Middle layer	8	1
Central atom for each block	0	8
	26	9

Now consider a cube consisting of three blocks in each direction, i.e. 27 blocks in all. Figure 1.12d shows a side view of this cube and the layers of atoms. This time there are two middle layers, and the count is

	Surface	**Interior**
Top and bottom layers	2×16	0
Middle layers	2×12	2×4
Central atom for each block	0	27
	56	35

These values have been entered into Table 1.1. A few more rows in the table could be completed in a similar way; but, as the number of blocks increases, the diagrams become increasingly tedious to draw! So the next step is to generalise from the examples involving small numbers of blocks to an algebraic formula. To do this, suppose a cube is made up of n blocks in each direction. By a process of logic, it is quite easy to work out formulae to fit the cases in Figure 1.12.

Consider the *surface atoms* first.

- The top and bottom layers each have $(n + 1)^2$ surface atoms.

- The 'middle layers' each have $4n$ surface atoms.

  ■ How many middle layers are there?

  ▨ For $n = 1$ there are none, for $n = 2$ there is one, for $n = 3$ there are two.

  ■ Generalising from these examples, how many middle layers are there, expressed in terms of n?

  ▨ For n blocks vertically, there are $(n - 1)$ middle layers.

Now consider the *interior atoms*.

- The $(n - 1)$ middle layers each have $(n - 1)^2$ interior atoms.

  ■ How many other interior atoms are there?

  ▨ There is one central atom per block. So for a cube of $n \times n \times n$ blocks there are n^3 'central' atoms in total.

You are now in a position to complete Table 1.1.

(i) By adding the various components above and simplifying the algebraic expression to the greatest possible extent, work out the formula for the total number, S, of surface atoms in a cube consisting of $n \times n \times n$ BCC-type blocks. [If you are not sure how to expand brackets or how to multiply two bracketed expressions, Box 1.4 shows the steps involved.]

(ii) By a similar process, work out the formula for the total number, I, of interior atoms. [*Hint:* in order to expand $(n-1)(n-1)^2$, begin by expanding the square, i.e. treat the expression as $(n-1)[(n-1)(n-1)]$ and work out the term within the square brackets first.]

(iii) Check that your formulae work by substituting $n = 3$ and making sure that you do obtain the values in Table 1.1. [If your values differ, check with the answer at the back of the book that you have worked out the formula correctly; once you have understood steps (i), (ii) and (iii), then carry on with the rest of the activity.]

(iv) Now complete all the blank cells in Table 1.1 (you may spot some symmetries that you can exploit). A few additional cells have already been filled in to help you. You do not need to work out the percentage of atoms on the surface to more than two significant figures.

(v) Iron has a unit cell edge length of about 0.3 nm. By referring to your completed table, calculate the maximum size of cube for which the majority of atoms are at the surface. In a cube of iron with edges 10 nm long, roughly what proportion of atoms are on the surface? How does this compare with the proportion in the bulk solid?

Table 1.1 Proportions of surface and interior atoms for a cube made up of $n \times n \times n$ BCC-type blocks

n	Number of blocks in cube ($n \times n \times n$)	Number of surface atoms	Number of interior atoms	Total number of atoms	Proportion of atoms on the surface
1	1	8	1	9	$8/9 = 89\%$
2	8	26	9	35	$26/35 = 74\%$
3	27	56	35	91	$56/91 = 62\%$
4					
5					
6					
33	3.6×10^4	6 536	68 705		
100	1.0×10^6	60 002	1 970 299		
1000	1.0×10^9		1 997 002 999		

You can check your completed table in the 'Comments on activities' section at the back of the book.

Box 1.4 Using brackets in algebra

An operation applied to an expression in a bracket must be applied to everything within the bracket. For example:

$(3a)^2 = 3^2 \times a^2 = 9a^2$

$(b + 2c) - (b + c) = b + 2c - b - c = c$

$(b + 2c) - (b - c) = b + 2c - b - (-c) = b + 2c - b + c = 3c$

$2(m + 3n) = 2m + 6n$

$2x(x - 2y) = (2x \times x) - (2x \times 2y) = 2x^2 - 4xy$

To multiply two bracketed expressions together, say $(a + b)$ and $(c + d)$, you need to multiply each term in the second bracket by each term in the first bracket, as illustrated by the red lines shown below. (Note that the multiplication sign between the first bracket and the second is implied and does not need to be written.)

$(a + b)(c + d)$

Multiplying the terms in order gives

$(a + b)(c + d) = ac + ad + bc + bd$

■ Rewrite the expression $(x + y)^2$ so that the brackets are removed.

$(x + y)^2 = (x + y)(x + y)$
$= x^2 + xy + yx + y^2$
$= x^2 + 2xy + y^2$

(since $xy = yx$). This process of turning a succinctly written expression such as $(x + y)^2$ into an equivalent, but longer, expression with no brackets is usually referred to as 'expanding' the brackets.

■ Rewrite the expression $(t - 2)^2$ so that the brackets are removed.

$(t - 2)^2 = (t - 2)(t - 2) = t^2 - 2t - 2t + 4 = t^2 - 4t + 4$

These kinds of calculation (which differ slightly for different atomic arrangements) explain the dominance of surface effects at the nanoscale. A nanoparticle has a far greater proportion of its atoms at the surface, and hence is far more reactive than a bulk solid. (Note that, given the size range you have explored in this activity, even a conventionally 'finely divided powder' would be considered a bulk solid.) Gold is a good example of the importance of scale in catalysis. Unlike bulk iron, bulk gold is a pretty unreactive metal and therefore makes a poor catalyst. Yet gold nanoparticles supported on a titanium dioxide surface have been found to make the most efficient catalyst for the oxidation of carbon monoxide into carbon dioxide:

$$2CO + O_2 \rightarrow 2CO_2$$

This is an important reaction, both commercially and environmentally: carbon monoxide is one of the products of incomplete combustion of petrol and is toxic. Catalytic converters in modern cars remove this and other undesirable gases from the exhaust. The normal catalysts are platinum and palladium–rhodium, all of which are expensive. The catalytic efficiency of nanostructured gold may make it viable in catalytic converters in the future. There turns out to be a very clear size dependence in the catalytic ability of the gold nanoparticles: maximum reactivity occurs with particles of about 3.5 nm in diameter. And reactivity is not the only property of gold nanoparticles that is affected by size, as you will see in the next two sections.

1.3.2 Melting point

The proportion of surface to interior atoms significantly affects the temperature at which a piece of solid substance melts. The melting process involves a supply of thermal energy that loosens the bonds between the atoms or molecules of the substance. If a solid is heated gently, it melts from the outside in (think of melting a lump of ice). So the tightness of binding between the atoms or molecules on the surface and those in the interior holds the key to the process.

Now think about how this process operates in the gold nanoparticles you considered at the end of Section 1.3.1. In a large (i.e. macroscopic) particle of gold, a given surface atom is tightly bound because of the many interior atoms exerting a pull on it. At the other end of the size spectrum, in a nanoparticle consisting of a few tens of unit cells, the ratio of surface to interior atoms is much greater and, therefore, a surface atom is much less tightly bound.

■   What does this imply about the relative energies required to melt a macroscopic lump and a tiny nanoparticle of gold?

▨   Less energy is needed in the case of the nanoparticle.

■   Given that thermal energy is directly proportional to temperature, what will be the relative melting points of the tiny piece of gold and the nanoparticle?

▨   The nanoparticle will melt at a lower temperature.

The melting temperature of nanoparticles below a certain size is indeed strongly dependent on size: the smaller the particle, the lower the melting point, as illustrated for gold in Figure 1.13. Notice that for the smallest size of nanoparticles shown on this graph, the melting point of gold is surprisingly low. (Remember that 300 K ≈ 27°C, the typical temperature of a hot summer day in the UK!)

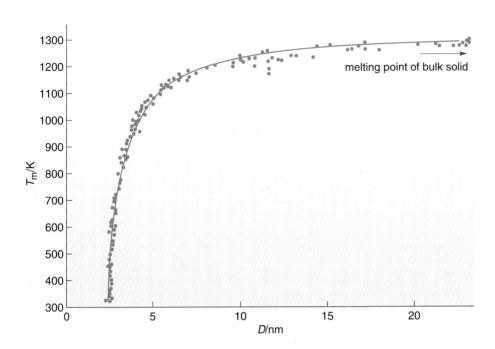

Figure 1.13 The melting temperature T_m of gold nanoparticles is dependent on the particle diameter D.

1.3.3 The colour of gold

Gold, silver and copper are all metals with such characteristic colours that their names are actually used to describe those colours. The wavelengths of light that a material absorbs determine its colour. Gold looks gold coloured when it reflects light because it absorbs all the incident light except light of those colours that when mixed together appear 'gold'. Nevertheless, it has been known since ancient times that very finely divided gold is not gold coloured at all, but red: there is a famous Roman cup in the British Museum in which gold particles have been incorporated into the glass. This appears red when light shines through it, but green in reflected light. Producing 'red gold' is a simple matter of chemistry. If a solution of a gold salt is treated with sodium citrate, then the product appears wine red. Michael Faraday (1791–1867) was the first person to suggest, in the 1850s, that this colour was due to finely divided metallic gold suspended in the solution. He even knew how to produce different colours by different chemical methods:

> the gold is reduced in exceedingly fine particles which becoming diffused, produce a beautiful fluid ... the various preparations of gold whether ruby, green, violet or blue ... consist of that substance in a metallic divided state.

> (Faraday, 1857)

The 'fluid' Faraday described is now known as a **colloid** (Box 1.5). This is a stable state intermediate between a solution and a suspension. Nowadays, it is possible to carry out high-resolution microscopic examination of the particles and to confirm that they are indeed tiny crystals of gold, no different in their atomic arrangement from bulk gold. Faraday could not explain why they are red or blue and not gold coloured, but we now know that quantum ideas hold the key to this phenomenon.

Box 1.5 Colloids

A colloid is a system in which extremely small particles (with sizes between 1 nm and 1 µm in any dimension) are uniformly dispersed through a second medium, which may be in the same or a different phase. The particles are not dissolved in the medium, nor are they in suspension: particles in a suspension would eventually settle out under gravity, whereas colloids are stable. Many well-known substances are colloids, as illustrated in Table 1.2. (The names used to classify colloids are included only for interest.)

Table 1.2   Classification and examples of colloids

Phase of dispersed particles	Phase of dispersing medium	Name	Example
liquid	gas	liquid aerosol	mist
solid	gas	solid aerosol	smoke
gas	liquid	foam	whipped cream
liquid	liquid	emulsion	milk
solid	liquid	sol	paint
gas	solid	solid foam	styrofoam
liquid	solid	gel	gelatin
solid	solid	solid sol	ruby glass

In a bulk metal, the electrons can move freely through the lattice, as described in Section 1.2.2. As a result, they can absorb photons of a wide range of energies; all that happens as a result is that the electrons take up that energy and move a little bit faster. In a nanoparticle, on the other hand, the electrons are confined in a very small space. You are probably already familiar with what happens when electrons are confined to the smallest possible spaces, i.e. the spaces inside atoms. Electrons in atoms have discrete energy levels and can, therefore, lose or gain energy only in specific amounts corresponding to transitions between the levels, giving rise to characteristic sharp lines in their emission and absorption spectra (Box 1.6). Electrons can absorb photons only of the right energy (i.e. of the correct frequency) that allow them to make transitions to higher energy levels. There are many more energy levels in even a small cluster of atoms than in a single atom, but the principle remains the same. Furthermore, because the exact number and spacing of the levels will depend on the number of atoms in a cluster, clusters of different sizes will absorb light of different wavelengths and will, therefore, have different colours. You will find out more about energy levels and the way they determine colours in Chapter 2.

Box 1.6 Revision of atomic energy levels

An atom can have only certain well-defined values of energy, represented as energy levels. Figure 1.14 shows the energy level diagram for hydrogen, in which the five lowest levels have been labelled E_1 to E_5. When the atom is in the lowest energy level of all, i.e. E_1, it is said to be in the ground state. To make the transition from level E_1 to E_2, which is of *higher* energy, the atom requires additional energy. It can acquire this energy by absorbing a passing photon (i.e. a particle of light or other electromagnetic radiation). If the energy of the photon E_{ph} is exactly matched to the energy difference between levels E_1 and E_2, then the atom can make the transition from level E_1 to E_2.

Transitions from lower to higher energy levels can take place between any two levels as a result of the absorption of a photon of the correctly matched energy (Figure 1.15a). Atoms can lose energy by the reverse process. If the atom makes a transition from a higher energy level to a lower one, then a photon is emitted, and this photon carries energy equal to that lost by the atom (Figure 1.15b). Each type of atom can be characterised by the energies of the photons it can emit or absorb and, therefore, has a unique 'spectral fingerprint'.

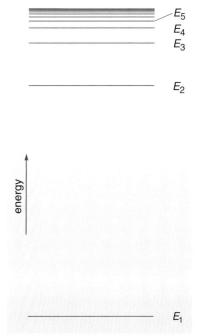

Figure 1.14 The energy level diagram of hydrogen. Energy increases from bottom to top. The differences between the energy levels represent the energies of photons that may be absorbed or emitted by hydrogen atoms.

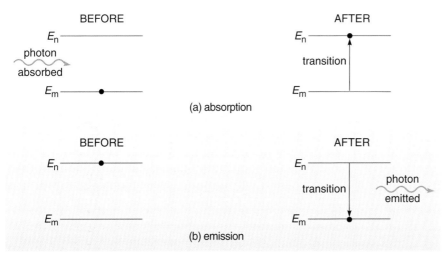

(a) absorption

(b) emission

Figure 1.15 (a) Absorption and (b) emission of a photon when an atom makes a transition between two energy levels E_m and E_n, such that $E_m < E_n$.

The energy E_{ph} of a single photon of electromagnetic radiation is proportional to the frequency f of the wave that characterises its propagation:

$$E_{ph} = hf$$

where h is a constant known as the Planck constant. The frequency scale of the electromagnetic spectrum is shown in Figure 1.16.

Figure 1.16 The electromagnetic spectrum, with the visible region of the spectrum shown in expanded form. Frequency f and wavelength λ are inversely proportional, related by the equation $c = f\lambda$, where c is the speed of light. Note that the wavelength is measured in metres. Frequency is measured in hertz (Hz), which are equivalent to cycles per second: $1\ \text{Hz} = 1\ \text{s}^{-1}$.

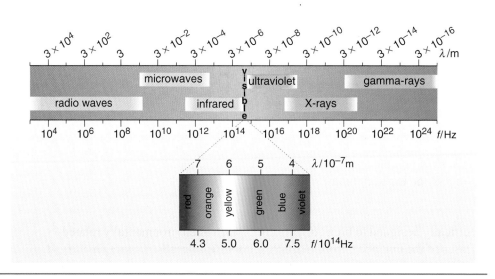

1.3.4 Clear sunblock

Particles not only absorb light, they also scatter it (i.e. reflect it in many directions). Bulk titanium dioxide (TiO_2) is a compound that scatters all colours of visible light very efficiently, so well in fact that under ordinary lighting conditions we see the sum of all the scattered wavelengths as white. For this reason, TiO_2 is widely used as a pigment and gives white paint its covering power. It has also been used for many years in high-factor sunblock, because it absorbs the UV radiation that damages the skin. Because of the light-scattering effect, this type of sunblock shows up white on the skin. As the TiO_2 particle size decreases, however, the scattering process becomes increasingly inefficient, while the UV absorption remains unaffected. TiO_2 nanoparticles of about 50 nm diameter are essentially 'invisible' to visible light: with wavelengths of between 300 and 700 nm, the light isn't scattered off such tiny particles. TiO_2 nanoparticles can, therefore, form the basis of a sunblock that cannot be seen on the skin but still offers the same UV protection as the old-style white variety. While some sportsmen have made a feature of their use of sunblock 'war paint' (Figure 1.17), non-visible sunscreen products are more acceptable to many consumers. For this and other similar reasons, the cosmetics industry has made a substantial investment in nanotechnology research. However, the incorporation of nanoparticles into cosmetics has raised various concerns about possible risks. In Section 1.3.5 you will start to consider how such risks may be evaluated.

Figure 1.17 Will nanotechnology make white sunblock a thing of the past?

1.3.5 Risk factors

In December 2003, the prestigious scientific journal *Nature* published its review of that year. The piece about nanotechnology was entitled 'What is there to fear from something so small?' As nanoparticles become more widely used in manufactured items – not just in catalysts and sunscreens, but also, as you will see in later chapters, incorporated into other types of material and pharmaceuticals – they will inevitably find their way into the environment and we will all come into contact with them. At the time of writing (2006), scientists do not actually know how much there is to fear from the release of particular kinds of nanoparticles. But they know enough to have concerns.

People have been exposed to tiny particles such as dust and smoke since the human species first appeared, and the body has many defence mechanisms to cope with them.

■   Suggest the three main ways in which small particles can enter the body.

▨   Particles may be ingested (swallowed) with food or drink, may enter through the skin – perhaps inadvertently via a cut or intentionally through injection – or may be inhaled (breathed in).

Of these three, ingestion of nanoparticles seems to pose the least risk. The gut is specifically designed to break down molecules, and environmentally related diseases of the gut, apart from reaction to bacteriologically contaminated food, are relatively rare. However, there has as yet been no real discussion of how, and in what concentration, nanoparticles might enter the food chain or the water cycle.

As far as absorption through the skin is concerned, the use of nanoparticles such as titanium dioxide (TiO_2) and zinc oxide (ZnO) in cosmetic products like sunscreens means that their toxicity must be seriously assessed. In particular, it is essential to find out whether the particles can penetrate the skin, and whether they then have the potential to generate **free radicals**. These are atomic or molecular species with unpaired electrons. For example, if a hydrogen molecule is broken down by photons of radiation, two free radicals are produced:

$$H_2 + E_{ph} \rightarrow 2H\bullet$$

The dot indicates an unpaired electron. If the radiation that provokes such a reaction is visible light, then the molecule is said to be photoactive. Free radicals are highly reactive, and may react with other species to produce other free radicals. For example:

$$2H\bullet + O_2 \rightarrow 2(OH\bullet)$$

■   Are the free radicals $H\bullet$ and $OH\bullet$ charged?

▨   No. $H\bullet$ is made up of one proton and one electron, so it is electrically neutral. An oxygen atom has six protons and six electrons, so the $OH\bullet$ free radical is also neutral, with a total of seven protons and seven electrons.

Free radicals can be produced only on the surface of particles, where the light strikes them. So, for a given mass of material, the smaller the particles (and hence the larger the surface area), the more free radicals can potentially be generated.

Free radicals are important in many essential biological processes. However, they can also react in undesirable ways that cause cell damage; it is thought that some cancers are the result of free radicals reacting with DNA. TiO_2 is photoactive, but the particles used in sunscreens are coated with silicates – primarily to prevent them agglomerating, but with the additional benefit of reducing the generation of free radicals. From those tests that have been carried out, it appears that the TiO_2 particles in sunscreens do not penetrate unbroken skin anyway. On the basis of this evidence, the Scientific Committee on Cosmetic and Non-food Products (SCCNFP) recommended to the European Commission in 2000 that TiO_2 be considered safe for use in cosmetics, whatever the size of the particles and whether or not they were coated. However, the Royal Society and Royal Academy of Engineering report published in 2004 notes that there have been few studies on the possible penetration by TiO_2 of *damaged* skin, for example skin already affected by severe sunburn. There is even more concern about ZnO, since there is evidence that 'microfine' (≤ 200 nm) particles do have a phototoxic effect on mammalian cells cultured in the laboratory. Further tests on living subjects were requested by the SCCNFP in 2003, and until they have been carried out a full safety assessment cannot be given.

When it comes to risks associated with the inhalation of nanoparticles, there is a wider range of studies on which to draw. Although specific studies of nanoparticles are still lacking, a lot can be inferred from what is already well known about what happens when other kinds of particles are inhaled. There is, for example, a great deal of accumulated evidence about the effect on workers in mining and quarrying of breathing in coal dust and quartz particles, and about the effect on the general population of atmospheric pollutants. In assessing the risks posed by nanoparticles in the air, the issues can be broken down into three categories:

1   What is the nature of the **hazard**? Hazard may be defined as the potential to cause harm. Examples of hazards here include the possibility of a 'cloud' of flammable nanoparticles catching fire or exploding, the potential of nanoparticles to generate free radicals that could damage cells, or the possibility of nanoparticles being toxic in other ways.

2   What is the **exposure**? Exposure is determined by the concentration of nanoparticles in the air, and the length of time that people spend breathing that air.

3   What is the **dose**? Dose may be defined as the number of nanoparticles that actually reach an organ where they present a hazard – the lungs in the case of inhalation. It is a function of both the exposure and the extent to which the body's natural defences eliminate particles before they actually reach the lungs.

The body has several mechanisms for removing potentially toxic particles that are inhaled. Some particles are simply deposited on the walls of the nose and throat, from where they are more likely to find their way to the gut than the lungs. If particles actually penetrate deep into the lungs, they are then attacked by **macrophages**, which are large cells that engulf and break down foreign particles, bacteria and general cell debris. Even if inhaled particles are not apparently toxic, it is believed that they may still cause lung disease if the dose is high enough, simply because the lung defences become overwhelmed. This is thought to be the situation with coal dust. Other small particles with low toxicity are in the air we all breathe. They are grouped together as 'particulates' and are typically of $10 - 100$ nm diameter. They are produced by combustion, as a result of both natural events (volcanoes and forest fires) and use of fossil fuels (power stations, vehicle exhausts, etc.). Many epidemiological studies of the effects of air pollution have shown it is exposure to particulates that is primarily responsible not only for lung disease, but also for asthma, heart disease and stroke in susceptible individuals. Particulates clearly present a hazard, even though their toxicity is low. Other particles are, on the other hand, extremely toxic: quartz particles of nanometre diameters are very reactive, and generate free radicals that damage defensive cells. Quartz particles have been shown to cause severe lung disease and lung tumours in rats.

It is not clear to what extent data relating to health risks from larger particles can be extrapolated to nanoparticles of the same chemical composition. For example, the effect of particle size on toxicity is not known, except that surface area is likely to be crucial because of the importance of surface activity in producing free radicals. Cellular uptake of particles also appears to depend on particle size. It is possible that the small size of nanoparticles may allow them to penetrate directly through cell membranes, and they may then disrupt cell functions. As one example of this, macrophages have evolved to take up larger particles, and nanoparticle penetration into these specialised cells could adversely affect their normal mechanism for breaking down bacteria. Until much more information is available, adherence to the precautionary principle seems a sensible course of action. Here, as with other aspects of risk assessment in nanotechnology, the primary focus should be on understanding the science involved. The global reinsurance company Swiss Re issued a report in May 2004 on the uncertainties and possible risks of nanotechnology. The title of this report says it all: *Nanotechnology – small matter, many unknowns.*

In this section we have concentrated on the effect of nanoparticles that may enter the body either through the local application of a cosmetic product or by inhalation. The exploration of potential new systems for delivering drugs means that medical and pharmaceutical researchers have also made studies of the effect of nanoparticles that enter the body through injection. These studies will be discussed in Chapter 3.

1.4 Nanoscale structures

This section introduces three types of structure that represent crucial steps in the development of nanoscience: carbon nanotubes, which are part of the fullerene family; supramolecular assemblies, which are made using one of the most important techniques of nanotechnology; and dendrimers, which are molecules with exceptionally high reactivity. In later chapters you will go on to consider some of the many applications of these structures.

1.4.1 Carbon nanotubes

In the early 1990s, fullerene research took an exciting new turn. By 1990, a method for mass-production of C_{60} and C_{70} had been established, as described in Box 1.1. The following year, Sumio Iijima, a Japanese scientist working for an electronics company, was experimenting with this technique when he discovered yet another form of carbon. Passing electric sparks between two graphite rods, Iijima vaporised the carbon and then condensed it as a sooty deposit. Examining the deposit using microscopy, he found not the expected ball-shaped fullerenes, but tiny tubes a few nanometres across, with curved caps at the ends. These **carbon nanotubes** immediately became the focus of intensive research in many laboratories. Iijima's original tubes were of the form now known as **multi-walled nanotubes (MWNTs)**, being hollow, with smaller ones concentrically nested inside larger ones, rather like Russian dolls (Figure 1.18a).

Within 2 years, Iijima and scientists at an IBM research facility in California were able to produce tubes with just one layer of carbon atoms, the so-called **single-walled nanotubes (SWNTs)**. The straight wall of an SWNT consists of a single layer of carbon atoms bound together in hexagons, like a layer of graphite rolled up to form a seamless tube. The ends of the tube are capped by a hemispherical fullerene-like structure in which carbon pentagons join to hexagons (Figure 1.18b).

Figure 1.18 (a) Schematic picture of an MWNT. (b) Part of an SWNT, showing how the individual carbon atoms are bonded.

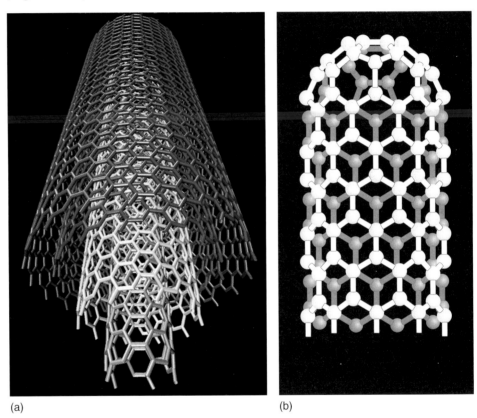

(a)   (b)

In Chapter 2 you will discover for yourself the different ways in which the 'rolling up' may take place and the different properties that the tubes can possess as a result. The diameter of an SWNT is 1.2–1.4 nm, and a typical length would be measured in micrometres. They are remarkably strong, but very light. However, high strength is just one of the properties of carbon nanotubes. The electrons within them are confined by the carbon monolayer to a space only about 1 nm across. This confinement gives rise to some very interesting electrical properties and quantum effects, as you will also see in Chapter 2.

Richard Smalley, who would soon be honoured with the Nobel Prize for his part in the discovery of C_{60}, switched his attention to carbon nanotubes shortly after they were characterised. In 1996, his group at Rice University devised the first method for producing bundles of SWNTs all aligned in the same direction to form a 'rope'. These ropes are typically 10–20 nm across and as much as 100 μm long. They have immense mechanical strength: if they could be grown to macroscopic lengths and wound together, then the resulting cable would be up to 100 times stronger than a steel hawser, but many times lighter.

As with nanoparticles, concerns have been expressed about health risks that might be associated with inhalation of nanotubes, especially because comparisons have been drawn with asbestos fibres, which are well known as a cause of several fatal types of lung disease and cancer. The hazards presented by asbestos are well understood, and depend crucially on the dimensions of the fibres: if they are less than about 3 μm in diameter they can reach the gas-exchanging tissues in the lung, and fibres above about 15 μm in length cannot easily be removed by macrophages (Figure 1.19). Nanotubes with diameters of a few nanometres and lengths in micrometres could pose a similar hazard, especially as they would probably be quite durable, so resisting attempts by the body's defences to break them down.

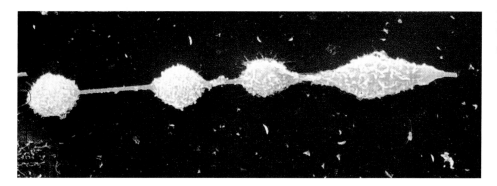

Figure 1.19 Four macrophages trying to ingest an asbestos fibre about 80 μm long.

Question I.2

Bearing in mind what you learned about hazard and exposure as contributors to the possible risk associated with nanoparticles, what kinds of study do you think are required in order to quantify the possible risks associated with inhalation of nanotubes, especially by workers involved in their manufacture, but also by the general public? (You should attempt this question and then check the answer before reading on.)

Methods of production currently result in nanotubes being made as agglomerates rather than as single tubes. They seem difficult to disperse, and it is thought to be unlikely that individual tubes would survive long in the air, as they would tend to clump together due to electrostatic attraction. However, at the time of writing (2006), their aerodynamic properties and toxicity are almost unknown. In addition, equipment required to measure them is highly sophisticated, expensive, and impractical for use in monitoring their concentration and size in the workplace or wider environment. Meanwhile, the huge potential of various kinds of nanotube for applications such as electronic devices indicates that production will substantially increase in the near future. The precautionary principle suggests that until the risks posed by such manufacture are quantified, nanotubes – like nanoparticles – should be treated as hazardous. The question of the possible introduction of a regulatory framework will be taken up in Chapter 4.

1.4.2 Supramolecular assemblies

The term 'supramolecular chemistry' was coined by Jean-Marie Lehn in 1978 to describe the study of chemistry that is literally 'beyond the molecule'. Lehn, together with Charles Pedersen and Donald Cram, was awarded the 1987 Nobel Prize in Chemistry for work in this field. Classical molecular chemistry is concerned with the ways in which atoms are held together by strong covalent bonds to form individual molecules; these molecules are typically in the 0.1–10 nm scale. In supramolecular chemistry, the building blocks are molecules, not atoms, and the scale of the assemblies is larger, typically 1–100 nm. The forces that are responsible for holding together molecular assemblies are weak and non-covalent: hydrogen bonding, London forces, etc. Lehn's definition of supramolecular chemistry as 'the designed chemistry of the intermolecular bond' emphasises the crucial importance of these weak interactions.

Two key concepts associated with supramolecular assemblies are **molecular recognition** and **self-assembly**, both of which span the divide between chemistry and biology. Molecular recognition, whereby one molecule in some way matches, and hence interacts with, another is exemplified by the way in which an enzyme matches with its substrate because of its shape, just as a lock and a key fit together (Figure 1.20).

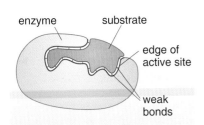

Self-assembly, the process whereby components spontaneously form ordered structures, is a fundamental principle of nature, occurring on all scales from the cosmological to the atomic. Thus, it is far from being unique to supramolecular systems, but it is at the molecular and biological end of the scale range that it has been most studied. In biological chemistry, self-assembly takes place both at the intramolecular level (protein folding is one example as you will read in Section 3.1) and at the intermolecular level. Self-assembly has long been discussed as offering a promising route for the development of nanoscale systems and machines; an extraordinary debate on this issue will be discussed in Chapter 4.

Figure 1.20 The lock-and-key model of enzyme action.

1.4.3 Dendrimers

Macromolecules are most commonly made from small molecules by a series of repeated reactions. You are probably familiar with the use of these kinds of reaction to produce long-chain polymers (Box 1.7). **Dendrimers** are polymers in which the monomer unit is branched. This branched structure is the origin of the name, from the Greek word *dendron*, meaning tree.

Box 1.7 Revision of long-chain polymers

Macromolecules that have a repeating structure because they are made from many identical small molecules are called polymers (from the Greek *poly*, many, and *meros*, part). The small reactant molecules are called monomers (from the Greek *mono*, single) and they bond covalently to one another to form the polymer; the repeating segment within the polymer molecule is called the monomer unit. In order to build up a long chain, each monomer must be able to form at least two covalent bonds to other monomers. Two types of polymerisation process can lead to the formation of long-chain molecules.

Addition polymers are formed by the reaction of monomers containing a C=C bond. One of the bonds within the double bond is broken and two new bonds to other monomers are formed; for example:

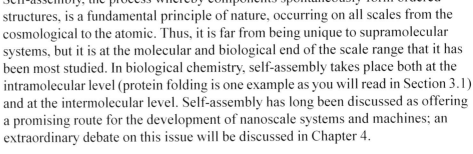

Here, X can represent a variety of possible atoms or groups. The chain can continue to grow by the addition of either monomers or other chains to either end of it.

■ What is the monomer unit in the above reaction?

▨ The monomer unit is:

This is the repeat unit within the polymer chain, with the dangling bonds corresponding to the bonds that are made when monomers are joined together.

Condensation polymers are produced when two monomers link up to form a larger molecule together with a molecule of water; for example:

This process can continue with other monomer molecules to build up the chain; alternatively, short chains can link together.

In both types of polymerisation, long chains may be formed by the reaction of shorter chains with one another or with monomers, so the chains harvested at the end of the reaction process usually have a range of lengths and molecular weights.

Dendrimers have several characteristic features that set them apart from long-chain polymers. First, and most importantly, they are constructed from branched monomers, so there is no backbone chain; instead, the polymer grows outward from the centre; for example:

This structure is one example of a PAMAM (PolyAMidoAMine) dendrimer.

Dendrimers of this type were the first to be synthesised, in the 1980s. Box 1.8 gives some background about amines and amides.

Box 1.8 Revision of amines and amides

Amines are molecules in which hydrocarbon groups (abbreviated as R) replace one or more of the hydrogen atoms in ammonia:

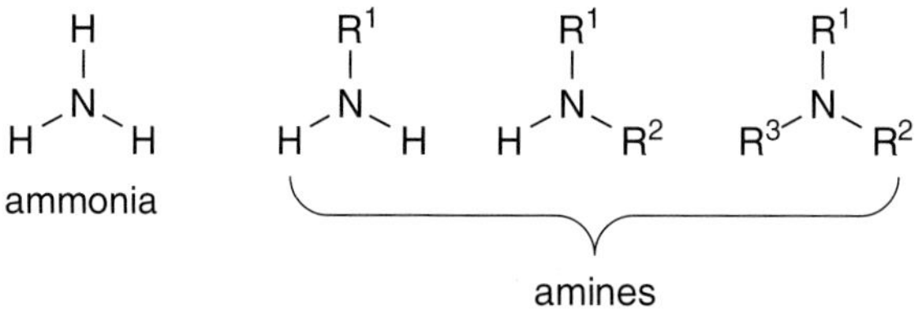

Amines with at least one hydrogen attached to the nitrogen react with carboxylic acids to give amides; for example:

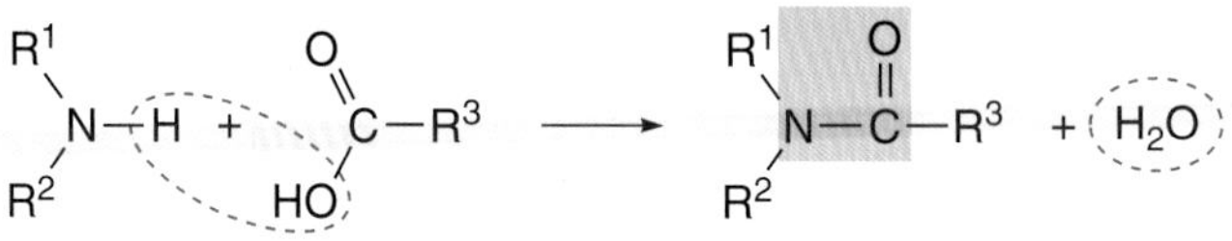

The carboxylic and amide groups are highlighted.

Amines can be used to make long-chain condensation polymers. Nylon is one example:

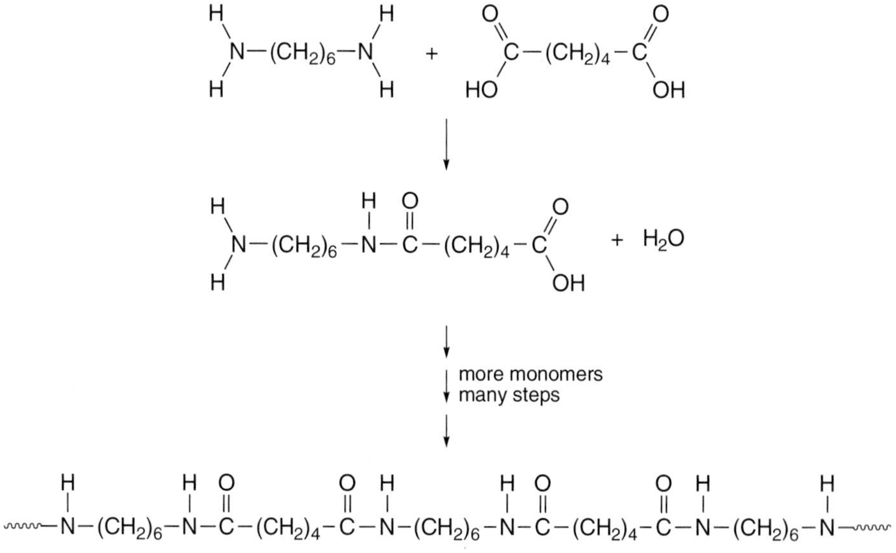

The starting point for the PAMAM dendrimer shown above is ammonia. Three branches emerge from the nitrogen 'core' at the centre of the molecule. The exact details of the synthesis are not important here (in fact, each monomer unit is added in two stages), nor is the composition of each monomer unit. The vital thing to appreciate is that each branch ends with an amino group (see margin)

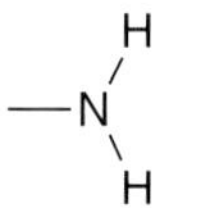

amino group

that can join with two new monomer units. This explains the second characteristic of dendrimers, namely that their synthesis proceeds in an iterative way. In each cycle, each of the reactive terminal groups on the edge of the dendrimer reacts with a monomer unit, so a complete new layer of branches is added. These successive layers are known as generations, as illustrated schematically in Figure 1.21.

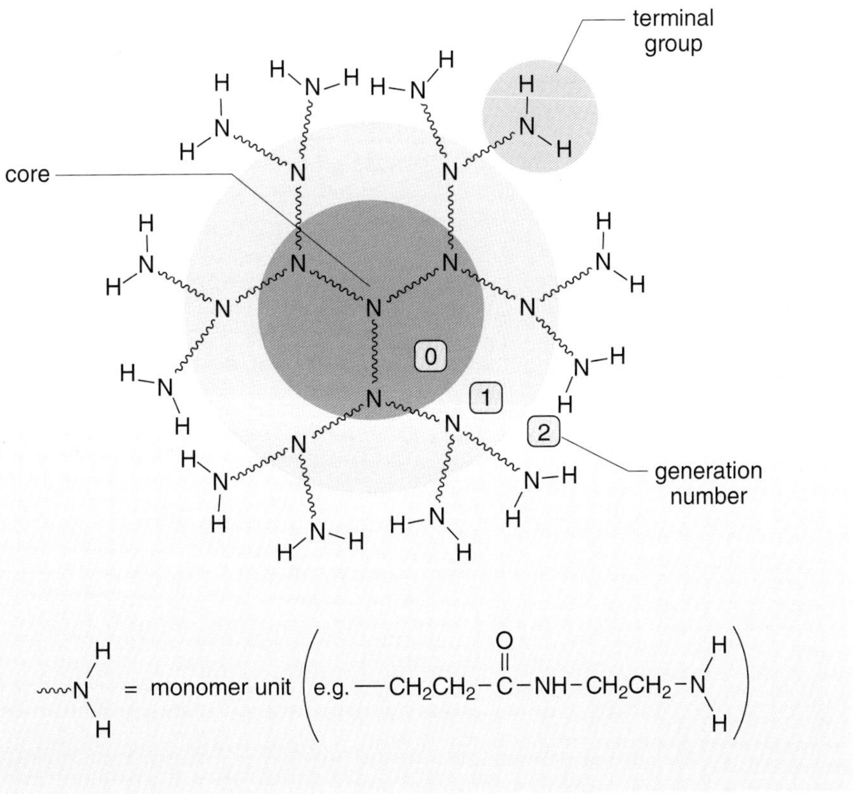

Figure 1.21 Generations of a PAMAM dendrimer based on ammonia (schematic representation).

It is quite easy to work out the number of reactive terminal groups for any given generation. It depends on:

(i)   The number of branches n emerging from the central core. If the starting point is ammonia, then $n = 3$, but other values are possible with other starter molecules.

(ii) The number of branches b at each junction. In the case of the dendrimer in Figure 1.21, $b = 2$.

(iii) The generation number G.

■ How many terminal groups Z are there in the zero, first and second generations for the dendrimer shown in Figure 1.21?

For $G = 0$, $Z = 3$

For $G = 1$, $Z = 6$

For $G = 2$, $Z = 12$

■ On this basis, work out the formula that links Z to n, b and G.

$Z = n \times b^G$

Table 1.3 shows the number of terminal groups by generation for the dendrimer in Figure 1.21.

Table 1.3 Number of terminal groups in a PAMAM dendrimer with an ammonia core and monomer units with two branches

generation	0	1	2	3	4	5	6	7	8	9	10
number of terminal groups	3	6	12	24	48	96	192	384	768	1536	3072

Table 1.3 shows how quickly the number of terminal groups increases with the generations. Low-generation dendrimers have quite flexible structures; but, as the number of generations increases, the branches on the periphery become more densely packed and the molecules develop a globular shape. Eventually there is no more room for further branches to be added and the polymerisation reaction effectively stops. For PAMAM dendrimers, this occurs after 10 generations, by which time the molecules are typically 10–15 nm across. This controlled type of synthesis means that dendrimers, unlike long-chain polymers, can be manufactured to high specification of size and molecular weight. Dendrimers fill the size gap between typical small organic molecules and the long-chain polymers, and this size, as well as their architecture, gives them some very useful properties. The uniform size can be used to calibrate molecular sieves. Their globular shape, coupled to their size, means they have a very high surface-to-volume ratio. In spite of their size, they are very soluble. This combination of large surface area and solubility lends itself to the use of dendrimers as 'carriers' for heterogeneous catalysis. The way in which the open, fan-like structure of early generations changes to a more densely packed globular structure in later generations results in a molecule with an outer 'skin' surrounding interior cavities (Figure 1.22). 'Guest' molecules can be introduced and held within these cavities. Drug delivery, about which you will find out more in Chapter 3, is just one possible application of this ability to host guest molecules within a soluble framework.

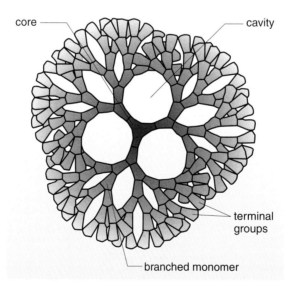

Figure 1.22 Schematic diagram of a dendrimer based on ammonia, showing cavities that could hold guest molecules.

There are many other types of dendrimer beyond the PAMAM variety; Figure 1.23 shows a few examples of the kinds of structure that have been achieved. The huge number of dendrimers patented in the early years of this century is evidence of the number of potential applications envisaged for these molecules.

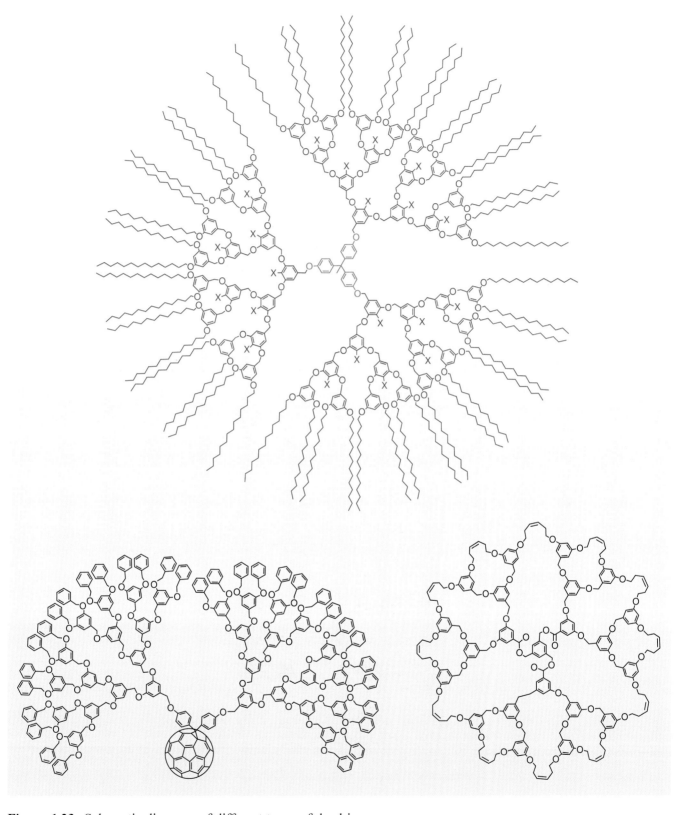

Figure 1.23 Schematic diagrams of different types of dendrimer.

1.5 Definitions revisited

We began this chapter with a basic working definition of nanoscale materials as those in which dimensions in the range 1–100 nm play a crucial role.

■ Do supramolecular assemblies, dendrimers and carbon nanotubes conform to this definition?

▧ Yes. Supramolecular assemblies and dendrimers have typical dimensions of 10–100 nm. Although carbon nanotubes can be up to hundreds of micrometres long, they are roughly 1 nm in diameter, and it is this small cross-section that conveys their special properties.

■ You have also met one important technique that operates below the 1 nm limit. What is this?

▧ Positional assembly: manipulation of individual atoms or molecules such as C_{60} (e.g. by an STM tip) is sub-1 nm science – atoms are typically ~ 0.1 nm in size.

One could, of course, simply revise the definition by extending the lower end of the length range downward by a factor of 10. A more useful approach, however, is to concentrate on the role played by the nanoscale on the properties of the materials. This not only gets away from classification problems at the extremes of whatever length range might be chosen, but also helps to distinguish nanoscience from more traditional areas of chemistry and molecular biology. This was the line taken by the Royal Society and Royal Academy of Engineering 2004 report, which adopted the following definition:

Nanoscience is the study of phenomena and manipulation of materials at the atomic, molecular and macromolecular scales, where properties differ significantly from those at larger scale.

■ What are the two main reasons described in this chapter for the fact that materials based on entities in the 0.1–100 nm range can have properties different from those on a larger scale?

▧ One reason is that, at a scale of order 10 nm, quantum effects become important (you have seen how these are exploited in the STM). The second reason is that a greater surface-to-volume ratio per unit mass – and hence the greater proportion of the atoms that are on the surface – compared with bulk materials results in increased chemical reactivity.

Nano*science* is concerned with studying these effects and the ways in which they affect the properties of materials. The aim of nano*technology* is to make use of the effects in order to produce functional materials and devices with novel properties. Because such production involves many different techniques and an enormous range of current and potential applications, the Royal Society and the Royal Academy of Engineering report refers not to an all-encompassing single field of nanotechnology, but to nanotechnologies in the plural, with the following definition:

Nanotechnologies are the design, characterisation, production and application of structures, devices and systems by controlling shape and size at nanometre scale.

What has been discussed in Chapter 1 is, broadly speaking, nanoscience. Chapters 2 and 3 discuss more of the science underlying some specific nanotechnologies and show how these are delivering new applications ranging from 'alternative energy' and nanoscale circuits to drug delivery.

Activity 1.2

Allow 45 minutes

To round off your work on this chapter, you should now watch the movie 'Nanotechnology', which you will find in the Topic 7 folder on the S250 DVD-ROM. In this documentary programme, part of a series entitled 'The Next Big Thing', experts in various scientific fields take part in a round-table discussion about their work in nanoscience and their hopes for future nanotechnologies. Two of the speakers (Harry Kroto and James [Jim] Gimzewski) have already featured in this chapter. You should also recognise many of the nanoscale structures and the techniques they describe, so the movie should help you to consolidate some of your knowledge and understanding of nanoscience. There are also some 'trailers' for further science to come in Chapters 2 and 3. However, do not be too concerned if you cannot follow every aspect of the speakers' discussions, especially their more speculative ideas about possible future developments.

The movie plays for 29 minutes, although you can pause it whenever you wish if you want to make notes or check back with parts of the text.

Summary of Chapter 1

Almost every branch of science now has specialists interested in nanoscale phenomena, and there are many areas of overlap in these interests.

The shapes and structures of carbon at the nanoscale (fullerenes and carbon nanotubes) differ from those of bulk carbon (graphite and diamond). Other nanoscale structures include supramolecular assemblies, which are held together by non-covalent bonds, and dendrimers, which are globular polymers. Nanoparticles and nanostructures have high surface-to-volume ratios, such that a substantial proportion of their atoms are on the surface. This can make them highly reactive and give them different physical properties from those of bulk materials. At a scale of ~10 nm, quantum effects become important, and this can also affect the properties of nanoscale materials.

The goal of nanotechnologies is to design and produce devices that put nanoscale properties to use. Techniques may be bottom-up or top-down; the former may involve positional assembly or self-assembly. Images of surfaces can be made with near-atomic resolution, and tools are available for moving and positioning individual atoms and molecules while simultaneously imaging the process.

The health risks associated with nanoparticles and nanotubes may be assessed by considering the hazard they present, the amount of exposure to the particles and the dose that may be received. Particles that can generate free radicals on their surface are likely to be particularly toxic. Nanoparticles may have the ability to penetrate cells through the membrane and may then adversely affect the functioning of the cell. Epidemiological studies of populations exposed to nanoscale particulates as a result of general air pollution have shown that even normally non-toxic substances may provoke toxic responses if they are inhaled at sufficiently high dose rate. No similar epidemiological data are available for carbon nanotubes, but extrapolation from the well-known harmful effects of inhaling asbestos fibres suggests that nanotubes might also be toxic in the lungs. Until risks can be assessed more accurately, adherence to the precautionary principle would require both nanoparticles and nanotubes to be treated as hazardous.

Questions for Chapter 1

Question 1.3

Figure 1.24a and b are schematic representations of an STM tip above a specimen that is not quite atomically smooth. Sketch onto these diagrams the path of the tip and the variation of the tunnelling current with position as the tip is scanned across the specimen (a) in constant-height mode and (b) in constant-current mode.

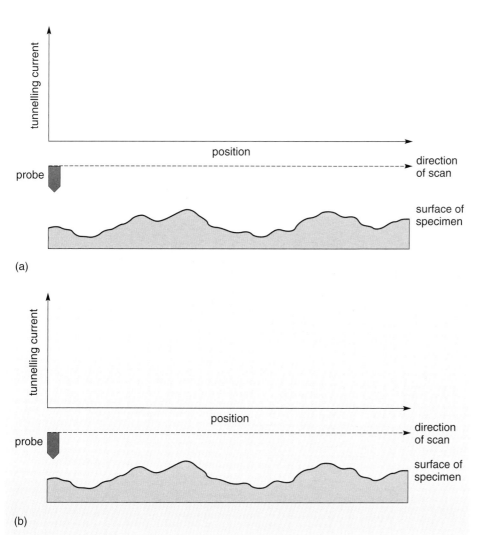

Figure 1.24 For use with Question 1.3.

Question 1.4

(a) Calculate the total surface area, volume and surface-to-volume ratio of a nanocrystal in the form of a perfect cube with sides 10 nm long. (b) What would be the radius of a spherical nanoparticle with the same volume? (c) If you were producing a nanoscale heterogeneous catalyst from a given mass of substance, which of these shapes would you choose?

(The surface area of a sphere of radius R is $4\pi R^2$ and its volume is $\frac{4}{3}\pi R^3$.)

Question 1.5

The report published in 2004 by the Royal Society and the Royal Academy of Engineering contains the following sentence: 'Free particles in the nanometre size range do raise health… concerns and their toxicology cannot be inferred from that of particles of the same chemical at larger size.' What size-dependent factors contribute to the possible health risks arising from free nanoparticles?

Question 1.6

Ammonia is not the only possible starter molecule for a PAMAM dendrimer. Another molecule that is commonly used is ethylenediamine:

$H_2N-CH_2-CH_2-NH_2$

core

monomer unit

(a) Sketch a schematic diagram similar to that in Figure 1.21, showing generations 0, 1 and 2 of a dendrimer based on an ethylenediamine core and a monomer unit that ends with an amino group. You may represent the core and the monomer unit as shown by the structures in the margin.

(b) How many terminal groups would this dendrimer have by the fifth generation?

Question 1.7

One of the drivers for nanotechnology noted in Section 1.1 is the hope that it will deliver smaller components for use in computer chips. Is it true, therefore, to say that miniaturisation is the principal benefit of nanotechnology?

Chapter 2

Nanotechnology and materials

In Chapter 1 you considered what it means to work at the nanoscale and looked in some detail at the tools that are available to help see what is going on at the atomic and molecular level.

This chapter will focus on one of the areas of nanotechnology that is developing very rapidly – that of materials. The overall properties of materials are determined by their structures at the atomic and molecular scale – as scientists understand materials at this level better, and their ability to control structures develops, so they have the potential to create a range of materials with very specific characteristics and applications.

The discussion will focus on a few of the main drivers for the interest in the development of nanomaterials. The first of these is to do with miniaturisation and electronics. As you saw in Chapter 1, Moore's law shows that the number of transistors on a computer chip has doubled roughly every 2 years for the last 40 years. Although Moore's law began life as an empirical rule, it has now become a driver for progress such that manufacturers feel compelled to achieve this doubling every 2 years. The public's demand for ever greater computing power has grown at least as fast, but it is clear that current miniaturisation techniques are approaching a level where nanoscale physical processes cannot be ignored. The next driver you will look at is concerned with energy production and environmental issues. There is a clear demand for methods of producing, or storing, energy that are cheap and non-polluting, and so there is much investment in proposed new technologies that aim to exploit the different physical properties of nanoscale materials to these ends. The chapter will end with a closer look at carbon nanotubes with regard to their unique physical properties and their potential use in a wide range of applications.

Many of the applications discussed are at an early stage of development, and in this chapter you will often be looking at potential or proposed technologies rather than existing ones. This is a different focus from Chapter 1, which was grounded very much in terms of already existing techniques. There are a number of questions you will encounter repeatedly which relate to the 'hype' surrounding nanotechnology and the extent to which existing science and technology research is being 'repackaged' in order to tap into the rapidly expanding funding opportunities for nanotechnology.

2.1 Electronics

As electronic components become ever smaller, there are three factors that need to be considered. These are (i) what can be physically manufactured, (ii) whether what has been made can be 'seen' and (iii) what are the physical properties of the thing that has been made. In Chapter 1 you revisited diffraction and saw how resolution is related to wavelength. This relationship partially defines the limits of the computer chips that can be produced. As the size of components becomes comparable with the wavelength of ultraviolet (UV) light, just beyond the visible part of the electromagnetic spectrum, they become more difficult to resolve.

■ In Chapter 1 you read that by 2005 the number of transistors per chip was set to exceed one thousand million. If one thousand million transistors are uniformly distributed on a square computer chip with sides of 1 cm, what is the *approximate* size of each transistor unit, to one significant figure?

▨ The chip area is $(1 \times 10^7 \text{ nm})^2 = 1 \times 10^{14} \text{ nm}^2$. If this area contains 1×10^9 transistors, then the area per transistor is $\dfrac{1 \times 10^{14} \text{ nm}^2}{1 \times 10^9} = 1 \times 10^5 \text{ nm}^2$ so each transistor unit has sides of length $\sqrt{10^5 \text{ nm}^2} \approx 300 \text{ nm}$.

■ At what size, in nanometres, do components become difficult to resolve by light with wavelength around 350 nm? This wavelength corresponds to the longest wavelength UV light, or the edge of the visible spectrum.

▨ Half of the wavelength is 175 nm, so structures of this size and smaller will be very difficult to resolve.

However, resolution is not the only issue as we get into the nanometre range. The patterning that defines a transistor has to be transferred using photographic and etching methods (more of which later) and it is getting very hard to make devices at this size. An even greater limitation is that, once components become smaller than around 100 nm, there are changes in physical behaviour as quantum effects begin to play a significant role. It is understanding, and hopefully exploiting, these effects that presents a fundamental challenge to the continuation of Moore's law.

This starts to set the scene for the challenges for electronics designers. Components cannot continue to get smaller using current technologies; but given the drive for ever greater computing power, scientists and technologists are now looking for completely new ways of designing computers that can exploit the different physical behaviour of components at the nanoscale.

2.1.1 Conductors, insulators and semiconductors

As we move on to look at examples of nanostructures, we will return again and again to the changes in properties that arise as we move from bulk materials to the nanoscale. You saw this in Chapter 1 in relation to some optical and thermal properties of gold, such as the melting point. In Chapter 1 you also revisited atomic energy levels. It turns out that the energy levels in solids determine many properties of the solid, most crucially their electrical properties. Other properties, such as the colours of some solids, are related to the energy levels of the atoms. In order to understand the changes that occur at the nanoscale it is important to be familiar with the basics of some of the bulk properties, in particular with energy levels in solids and how these affect electrical properties.

In Chapter 1 you learned that metals are electrically conducting because they contain 'free', or valence, electrons. Insulators, by contrast, have all their electrons bound more closely to the nuclei of the parent atoms, such that the electrons are not able to move easily through the material. In between these two is a small group of materials called **semiconductors**, such as germanium and silicon. These are neither good conductors nor good insulators. In their pure crystalline state they are quite good insulators when kept very cold, but their conductivity increases enormously when even as few as one atom in 10 million is replaced with an impurity that adds or removes electrons from the crystalline structure. This addition of a very small proportion of foreign atoms is called **doping**. For example, if a phosphorus atom, which has five valence electrons, is introduced into a lattice of silicon, which has four valence electrons, then there is a 'spare' electron that is only loosely bound to the impurity atom and which can be thermally excited and can move under the influence of a battery. Materials that have been doped in this way can be made to behave sometimes as insulators and sometimes as conductors.

There are two basic types of semiconductor: n-type (negative) and p-type (positive). In an n-type semiconductor, the carriers of current are free electrons, as in a metal. Semiconductors differ from metals in that their ability to conduct electricity, or **conductivity**, increases with temperature, whereas the opposite is true for a metal. In a metal, the electrons are not tightly bound to their parent atoms, but form an electron gas, as discussed in Section 1.2.2. As you saw in that section, the electrons are able to move easily, which is why metals are generally good electrical conductors. However, the lattice that the electron gas moves around is not perfect: where there are defects, electrons will scatter off the imperfections and this scattering is a friction-like effect. The more energy the electrons have, the more easily they are scattered and the greater the friction they feel. Hence the conductivity falls as the temperature rises. In a semiconductor, however, the extra energy provided by the increase in temperature shakes electrons loose from the atoms of the semiconductor and because the conductivity of a semiconductor depends on a small number of charge-carriers, unlike a metal, the extra free electrons more than compensate for the extra friction they feel at the higher temperature. Hence the conductivity of a semiconductor rises as the temperature rises.

In a p-type semiconductor, the current carriers are **holes**. To understand what this means, consider a row of neutral atoms as shown in Figure 2.1a. If an electron is pulled out of the atom at the right-hand side by a battery, for example, there is a hole, or missing electron at that position, as in Figure 2.1b. Figure 2.1c shows how an electron from the next atom along will jump into this hole, leaving in turn a hole at the position of the second atom. This continues and the movement of electrons from left to right can be described as the movement of a hole from right to left, carrying a positive charge with it. This is basically how conduction occurs in a p-type semiconductor. Instead of a gas of free electrons, there is a gas of free holes. A hole is just like a 'positive electron'; the reason it is discussed in terms of the 'absence' of an electron is that atoms are composed of neutrons, protons and electrons – the hole is not a fundamental particle.

The concentration of free electrons and free holes is largely determined by the impurities that are present in the material. **Donor** impurities consist of atoms that release their valence electrons when placed in the semiconductor, thereby increasing the number of free electrons. **Acceptor** impurities consist of atoms that trap electrons when placed in the semiconductor, thereby generating holes.

■   What types of impurity lead to each of n-type and p-type semiconductors?

▨   A semiconductor with donor impurities is n-type and one with acceptor impurities is p-type.

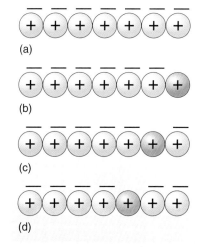

Figure 2.1   A row of neutral atoms, composed of positive ions (balls marked with +) and electrons (marked with −) to illustrate the movement of holes in a semiconductor.

You will be familiar with the idea of energy levels in individual atoms from Chapter 1. Any type of atom can absorb or emit only photons of specific energies in order to make transitions between quantized energy levels, as you saw in Box 1.6.

When a solid is formed and there are many atoms together, some of the electrons are shared by neighbouring atoms, and in the case of metals, by the whole array of atoms. Each atom in the solid still has energy levels associated with it, but there are many more energy levels than there would be for an isolated atom and they are generally so close together that they form a continuous energy band, typically a few electronvolts wide (Box 2.1).

Box 2.1 Revision of electronvolts

It is convenient when dealing with energies on the scale of light photons to express them in a unit called the electronvolt, which has the symbol eV. One electronvolt is the energy converted when one electron moves through a potential difference, or voltage, of 1 V. When a charge Q moves through a voltage difference ΔV, the energy transfer ΔE, in joules, is the number of coulombs of charge transferred times the voltage difference, i.e. $\Delta E = Q\Delta V$. The magnitude of the charge on one electron is 1.6×10^{-19} C. So:

$$1 \text{ eV} = (1.6 \times 10^{-19} \text{ C}) \times (1 \text{ V})$$

$$= 1.6 \times 10^{-19} \text{ J}$$

■   A photon of yellow light has an energy of about 3.5×10^{-19} J. What is this value expressed in electronvolts?

▨   Since $1 \text{ eV} = 1.6 \times 10^{-19}$ J, then

$$1 \text{ J} = \frac{1 \text{ eV}}{1.6 \times 10^{-19}}$$

So the energy of this photon of yellow light

$$E_{ph} = 3.5 \times 10^{-19} \text{ J}$$

$$= 3.5 \times 10^{-19} \times \frac{1 \text{ eV}}{1.6 \times 10^{-19}}$$

$$= 2.2 \text{ eV}.$$

Continuous energy bands are not the whole story, however; within these energy bands there are gaps where no energy levels are found. These gaps are usually called **band gaps** or energy gaps. Electrons can have energy values that exist within one of the bands but cannot have energies corresponding to values in the gaps. This is illustrated in Figure 2.2.

It is impossible to explain the band structure of solids without some advanced quantum mechanics. However, in a very hand-waving way, you can think about what happens when atoms that were well separated are brought close together. The states that defined electrons in individual atoms begin to overlap. Electrons can then move from one atom to another and, therefore, not be identified with a particular atom. The states that were associated with individual atoms become

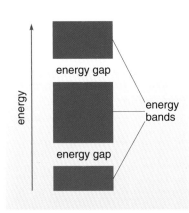

Figure 2.2   In solids, energy levels are concentrated into bands. These bands are separated by energy gaps in which there are no energy levels.

changed to states stretching over all the atoms. Many different such states arise and all have different energies, such that an atomic energy level broadens into a band of energy levels.

The lowest energy states are due to the inner atomic levels and these bands are full with electrons. These electrons are bound tightly to the nucleus and are unable to change their state and so do not form chemical bonds with other atoms. The outer electrons of the atoms, which are the ones responsible for chemical bonding, form the **valence band** of the solid.

It is the band structure that accounts for the electrical properties of solids. In order to take part in the conduction of electricity, an electron must be able to change its energy by small amounts, gaining and losing energy as it moves through the solid, and so moving from one energy level to another. An electron can do this only if there are spaces in energy levels that are energetically available to it. In general, if the valence band is full, then electrons cannot change to different energy states in the same band. For conduction to occur, electrons have to move to an unfilled band, i.e. the **conduction band**.

A conductor is characterised by having a significant number of electrons in a partly filled conduction band. This means that there are a large number of nearby empty levels, or states, into which these electrons can move if electrically or thermally excited.

For a material which is an insulator, the valence band is full of electrons that cannot change their energy. The conduction band is far above the valence band in energy and a negligible number of electrons can be excited out of the valence band and into the conduction band.

In the case of a semiconductor, the gap between the valence and conduction bands is much less and the heat content of the material at room temperature can thermally excite some electrons from the top of the valence band to the bottom of the conduction band where they can take part in conduction. The number of electrons reaching the conduction band by this thermal excitation is relatively low, but it is not negligible, and so the electrical conductivity is small. A material of this type is called an intrinsic semiconductor. Illustrations of energy bands for conductors, semiconductors and insulators are shown in Figure 2.3.

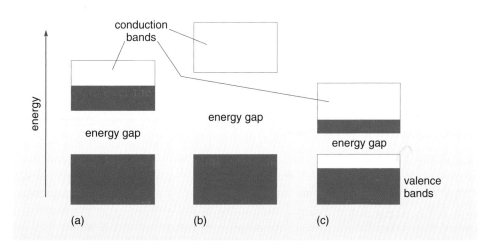

Figure 2.3 Energy bands of (a) a conductor, (b) an insulator and (c) a semiconductor. The dark-blue shading indicates the presence of electrons, the pale-blue shading the absence of electrons.

When intrinsic semiconductors are doped with donor atoms that give electrons to the conduction band, current can be carried more easily. The donor atoms have energy levels just below the conduction band of the semiconductor, so that electrons can be easily excited into the conduction band. Similarly, semiconductors can be doped with acceptor atoms which have energy levels just above the valence band. In this case the acceptor atoms accept electrons from the valence band, leaving behind holes that also carry current, as you have seen. The first of these processes produces n-type conductivity and the second produces p-type conductivity. These acceptor and donor energy levels are shown in Figure 2.4.

Figure 2.4   The energy gap of a semiconductor showing acceptor levels above the top of the valence band and donor levels below the bottom of the conduction band.

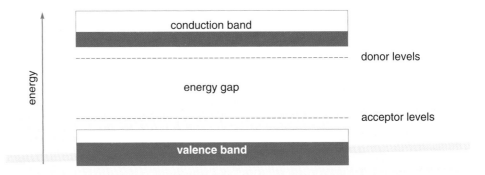

2.1.2 Electronic components and circuits

Electronic circuits involve three levels of organisation. Among the basic building blocks are transistors and diodes, which act as switches that can turn current on and off, control its flow, and amplify signals. At the next level are the interconnecting wires that link the components together to perform operations. The third level is the overarching way the components are connected so that the circuit can operate as a system. This chapter won't consider the system level of circuitry in much detail. However, before looking at the developments in terms of nanotechnology, it is useful to consider briefly how electronic circuits currently work and are designed.

A **diode** is a semiconductor crystal which has been doped in such a way that one-half of it is p-type and the other half is n-type. The details of how diodes work are beyond the scope of this course but, essentially, one-half is easy for electrons to move through and the other is difficult. It is a bit like swimming through water on one side and through mud on the other. If electrons starting on the 'mud' side are given enough energy to make it through the mud they can easily get all the way across. On the other hand, if they start on the 'water' side and are only given enough energy to get through that, they won't be able to get through the mud when they reach it. Hence, a diode is a device through which electric current can flow in only one direction.

Transistors are also semiconductor devices. There are a variety of different types of transistor and, again, the details of their function are beyond this course. In simple transistors the semiconductor crystal has three parts – either n–p–n or p–n–p types. In the first of these the electrons are the charge carriers and in the second the holes carry the charge, but they are otherwise very similar. As with the diodes, exploiting the boundaries between the electron-rich n-type semiconductor and the hole-rich p-type and applying varying voltages to the three sections allows the current flow to be controlled such that transistors act as switches and amplifiers of electrical signals. Other transistors exploit the fact that the amount of current that flows through a semiconductor channel can be varied by applying a variable voltage; this allows them to act as switches and amplifiers. As you have seen, a modern computer chip contains millions of transistors.

Integrated circuits, such as are used for computer chips, are electrical circuits with all the transistors, diodes, wires and other devices built into a small square of the semiconductor silicon. They are made by a process known as **lithography**. Lithography is a printing technique. Originally developed in the 1790s, it is a method of printing from a flat surface whereby the design would be made on the surface from a greasy, water-repelling material so that water applied to the surface is absorbed where there is no design. Oil-based ink, then, sticks to the design, but not to the moist areas. Pressing paper onto the surface produces a print. This is also rather like screen-printing, where a waxy pattern on a screen allows ink through to produce an image of the pattern.

The standard fabrication method used to make microelectronic devices is photolithography. In a similar way to the printing technique, this uses light-activated chemistry to define a pattern which can be transferred to a **substrate,** which is a silicon wafer, or thin slice, on which electronic devices are created. Polymers called photoresists are spread in a layer, a few nanometres thick, on the substrate and exposed to UV light. This produces chemical changes such that either the exposed or the unexposed parts of the polymer layer can be dissolved and washed away in some kind of developer. However, there are other lithographic methods that are being used to produce nanoscale patterns. The most common of these is electron-beam lithography (EBL). Exposure to an electron beam alters the chemistry of a resist layer on a semiconductor substrate. EBL can routinely produce lines as narrow as 20–30 nm. The biggest problem with EBL is speed. In photolithography a whole substrate layer can be exposed at one time to UV light, whereas EBL requires the electron beam to draw each feature one at a time. It is an ideal technique for producing small patterns, but it is not cost-effective for large-scale production.

Scanning probe lithography uses the STM and AFM that you met in Chapter 1. There are several techniques, which differ in their detail but all of which allow very precise structures to be created by moving individual atoms and molecules across substrate surfaces. It is very unlikely, given the number of atoms and molecules that would need positioning, that these techniques will ever be suitable for mass production.

2.1.3 Down to the quantum level

You have already seen in Chapter 1 that physical properties can change as the size of a material is reduced to the nanoscale. You were introduced to nanoparticles and nanotubes with dimensions at this scale, and what you learned there about their properties applies here. In developing nanoscale wires and components for electronic circuits it is important to understand what physical changes occur as a component's dimensions are reduced to the nanoscale. If two dimensions are reduced to this scale, with one remaining comparatively large, we obtain a *quantum wire* and if all three dimensions of a semiconductor component are reduced into the nanometre range then we obtain a *quantum dot*. The key word here, as you might have guessed, is quantum.

Confinement means that particles' motions become limited by the physical size of the region in which they are moving. In general, in electronic systems, the wire and component dimensions are very large compared with the distances between atoms and the conduction electrons are delocalised, i.e. they are able to move freely around throughout the conducting medium. However, when one or more dimensions are reduced to just a few times greater than the separation between the atoms in the lattice, the delocalisation of the electrons is impeded and they experience confinement. The influence of electrostatic, or Coulomb, forces becomes more pronounced (Box 2.2 explains how the force gets greater as the separation between charges gets smaller) and the electrons' motion becomes restricted. This is what happens in atoms, and the limitations on movement give rise to the atomic energy levels discussed earlier in Box 1.6.

Box 2.2 Revision of Coulomb's Law

Two particles of unlike (or like) charge, Q_1 and Q_2, at rest, separated by a distance r, attract (or repel) each other with an electric force, F_e, that is proportional to the product of their charges and inversely proportional to the square of their separation :

$$F_e = -k_e \frac{Q_1 Q_2}{r^2} \text{ where } k_e \text{ is a constant.}$$

Confinement at the nanoscale is not exactly the same as confinement in an atom. Even in one dimension there will be several atoms in a nanoscale layer. Consequently, there are more energy levels than in atoms. You have seen that in a solid, where the electrons are delocalised, there are so many energy levels that they form a continuum and in an atom there are sharply defined discrete energy levels. In quantum layers, wires and dots, the situation is somewhere in between, with comparatively few energy levels that are well defined. The number of atoms, the types of atom and the prevailing conditions can be varied and 'tuned' to control the size of the gaps between energy levels, and thus the energies of the photons that may be emitted or absorbed.

So, if nanoelectronic circuits are to become a reality, it is clear that it will not be enough to be able to miniaturise the types of component and connections that exist in microelectronics. It will be essential to understand, and utilise, the quantum nature of the components and wires that will necessarily be involved.

2.1.4 Quantum dots

Quantum dots are nanoparticles of semiconductor. They are sometimes called 'artificial atoms' because, as you have seen in the Section 2.1.3, the energy levels in a nanoscale dot are relatively few and sharply separated in energy, rather like the energy levels of an atom. You will see in this section that the wavelength of light (which determines its colour, see Figure 1.16) emitted by these quantum dots under UV illumination depends very sensitively on the size of the dot. Being able to produce and control the colour that the dots emit makes them potentially very useful for colour displays, like computer screens, or as markers in biological processes. You will read more about these later in this and in the following chapter. Ultraviolet illumination is used, as it is composed of higher energy photons than visible light and so provides enough energy to excite the electrons in the nanoparticles to emit a full range of colours. By varying their size, particles can be made to emit or absorb specific wavelengths of light.

■ Given that visible light has wavelengths ranging from about 400 to 700 nm, what frequency range does this correspond to? Refer back to Box 1.6 for the relationship between frequency, wavelength and the speed of light, c. (Remember $c = 3 \times 10^8$ m s^{-1}.)

▨ Recalling the relationship $c = f\lambda$, this range of wavelengths corresponds

to a frequency range between $f = \dfrac{c}{\lambda} = \dfrac{3 \times 10^8 \, \text{ms}^{-1}}{4 \times 10^{-7} \, \text{m}} = 7.5 \times 10^{14} \, \text{Hz}$ and

$$f = \dfrac{c}{\lambda} = \dfrac{3 \times 10^8 \, \text{m s}^{-1}}{7 \times 10^{-7} \, \text{m}} = 4.3 \times 10^{14} \, \text{Hz}$$

To understand how the size of quantum dots affects their light emission, you need to recall that a semiconductor is characterised by a band gap between its valence band and its conduction band. Remember also that the bands are really a continuous series of energy levels due to the atoms in the crystal structure.

Suppose the size of the semiconductor crystal is reduced. As each atom is removed, its contributions to both the valence and conduction bands are also taken away. As the crystal reduces in size, two things happen. First, as you have seen in Section 2.1.3, each band ceases to be a continuum of energy levels and individual levels are revealed. Second, the easiest atoms to remove are those at the band edges, where there were fewest to start with; and removing these increases the size of the band gap.

So we have a quantum dot. 'Quantum' because the energy levels are no longer a continuum, but are discrete or quantised, and 'dot' because these effects are only seen when the crystal is very small. Removing more atoms increases this effect: as the dot gets smaller, so the band gap gets bigger, as you can see in Figure 2.5a.

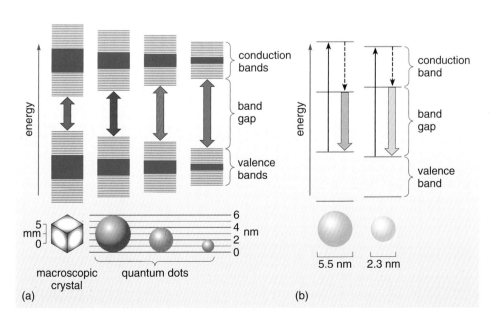

Figure 2.5 (a) A semiconductor is characterised by a band gap between its valence band and its conduction band. As the semiconductor crystal is reduced in size, each band ceases to be a continuum of energy levels and individual levels are revealed. At the same time, the easiest atoms to remove from the crystal are those at the band edges and their removal leads to an increase in the band gap. (b) Two different coloured quantum dots can be excited by the same light source but still emit their own characteristic colour photons.

Now consider two different-sized dots, as shown in Figure 2.5b. When UV light shines on the dots, electrons can absorb photons of energy which are enough to excite them out of the valence band and into the conduction band. The energy required to promote an electron to the top of the conduction band is about the same for both dots, although their band gaps are quite different. This means that both dots can be excited by the same light source.

The electrons then fall, in small steps, to the lowest energy level in the conduction band. These small steps correspond to energies in the infrared (IR) part of the spectrum and so small amounts of heat are emitted (as shown by the broken arrows in the figure) but no photons in the visible part of the spectrum are emitted. Once at the bottom of the conduction band they then fall back to the top of the valence band, and this larger gap results in its emitting its excess energy as a photon of visible light. Since the band gap for the larger dot is smaller, the larger dot emits a less energetic photon than the smaller dot.

Since the energy of a photon is inversely proportional to its wavelength, the larger dot will emit light of a longer wavelength (i.e. towards the red end of the spectrum) than the smaller dot, whose photon will be further towards the blue end of the visible spectrum. The different colour emissions that can be obtained with quantum dots are illustrated in Figure 2.6.

Figure 2.6   Quantum dots of the same material but different sizes (here, cadmium selenide in suspension) have different band gaps and emit different colours.

Fabricating quantum dots of exactly the sizes required is not simple. Nonetheless, there is sufficient belief in their usefulness in electronics that a variety of different fabrication techniques are being explored and developed. One is the top-down lithographic approach of using a radiation-sensitive resist on a semiconductor substrate and etching away the unwanted material, leaving quantum dot 'islands' on the substrate. However, as you saw before, such techniques are relatively slow and production of commercially viable quantities of quantum dots by these methods is unlikely. The other problem is that etching away the resist means discarding more than 90% of the material. It is usually more desirable to import 10% than to export 90%. The approach which is finding greater application is thus a bottom-up chemical synthesis method where a sequence of carefully controlled chemical reactions leads to gram quantities of quantum dots being prepared.

Question 2.1

How many quantum dots are there in a gram? This is not an entirely straightforward calculation and needs to be done in a 'back of the envelope' way. This means you need to think about orders of magnitude rather than exact numbers, but be aware of any assumptions or rounding errors you introduce.

Use the following information to estimate the number of quantum dots of silicon in a gram:

- the average diameter of a quantum dot is occupied by approximately 50 atoms;

- the atomic mass of silicon is 28 – recall that this means Avogadro's number (6.02×10^{23}) of silicon atoms has a mass of 28 g;

[*Hint*: First estimate the number of atoms in a quantum dot. Remember that the volume of a sphere of radius R is given by $\frac{4}{3} \pi R^3$.]

How might quantum dots be used? Probably their most advanced development in terms of electronics is in light emitting diodes (LEDs). LEDs are a type of semiconducting diode, which is a device that has two regions of differently doped semiconductor that are close together. The n-type region, you should recall, has many electrons and the p-type region has a shortage of electrons or an abundance of holes. Between the n- and p-type regions is a narrow layer, called the **depletion zone**, which is devoid of both electrons and holes. On connecting the negative terminal of a battery to the n-type region and the positive terminal to the p-type, the electrons and holes are driven into the depletion zone where they can meet. When an electron and a hole meet, they annihilate each other. Since a hole can be described as the 'absence of an electron', when an electron meets a hole it falls into it and both cease to exist. This process of annihilation does lead to a net release of energy though, and photons are produced. It is this process that is at work in LEDs. LEDs are common in our lives. The advantage of quantum dot LEDs is that the size of the semiconductor nanoparticles will determine the colour of light emitted – and the light is quite pure, so that, for example, green light emitted is relatively uncontaminated by yellow or blue light. This improves clarity if LEDs are used as pixels in colour displays, such as for computer screens or televisions.

Quantum dots are also finding uses in areas other than electronics. One of the most promising applications is to use quantum dots instead of fluorescent dyes for monitoring the movement of cells and biological molecules. One light source can be used to stimulate a range of dots; and since the light they emit is pure, being of a single wavelength, differently coloured dots can be used to track different biological processes without their emissions interfering with one another. You will read more about these applications in Chapter 3.

2.1.5 The case of Hendrik Schön

Earlier in this chapter you read quite a lot about electronic components. The ability to manufacture ever-smaller transistors has been highlighted as one of the key drivers in electronics development. Transistors are basically electronically controlled switches and are currently made from silicon. Silicon has the right semiconductor properties and, just as importantly, we know how to manufacture silicon transistors quickly and cheaply by the million. But there are some properties we might like in a transistor that silicon does not have. The most important of these is flexibility. If transistors can be made from organic molecules with the required band structure, then they would be more flexible – offering, one day, the possibility of roll-up computer screens, or electronic fabrics or plastics. In 1998, Bell Laboratories in the USA employed Hendrik Schön as a postdoctoral researcher. He was 28 years old at that time and in his 4-year career at Bell Laboratories he made a series of spectacular breakthroughs in the fields of organic electronics, superconductivity and nanotechnology. By the beginning of 2002 Schön had produced more than 90 papers. In 2001 he churned out a paper every 8 days on average, several of which were published in *Science* and *Nature*. It is very rare to get so many papers in such prestigious journals, and also the writing and peer-review processes tend to take time. Schön was getting a lot of attention in the scientific community, winning prizes and being tipped as a possible future 'great' scientist. His work during this period culminated in the announcement that he had produced a transistor made from a single organic molecule. This extract from a Lucent Technologies (who own Bell Laboratories) press release in November 2001 shows how highly this development was valued. The press release was entitled 'Bell Labs scientists build the world's smallest transistor, paving the way for "nanoelectronics"':

> Scientists have been looking for alternatives to conventional silicon electronics for many years because they anticipate that the continuing miniaturization of silicon-based integrated circuits will peter out in approximately a decade as fundamental physical limits are reached. Some of this research has been aimed at producing molecular-scale transistors, in which single molecules are responsible for the transistor action – switching and amplifying electrical signals.

> Bell Labs' 'nanotransistors' – so-called because they are approximately a nanometer, or one-billionth of a meter, in size – appear to rival conventional silicon transistors in performance. They are made using a class of organic (carbon-based) semiconductor material known as thiols. In addition to carbon, thiols contain hydrogen and sulfur.

> The main challenges in making nanotransistors are fabricating electrodes that are separated by only a few molecules and attaching electrical contacts to the tiny devices. The Bell Labs researchers were able to overcome these hurdles by using a self-assembly technique and a clever design.

In early 2002, questions began to be raised around Schön's work. Other researchers could not seem to reproduce his results, which was unusual, and then colleagues noted that figures in two of his papers appeared identical, even though they represented measurements done at different temperatures, and temperature should have affected the results. A number of incidences of this kind of duplication were found once colleagues began to look into it, and Figure 2.7 shows an example of results from three different papers. Not only are the graphs, shown on the left of the figure, very similar, but the noise, which is magnified in the main part of the figure, is also identical. Noise is generated by random background processes – it is virtually inconceivable to produce more than one set of results with the same background noise.

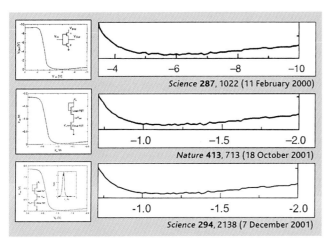

Figure 2.7 Schön's published data from studies of different devices revealed a similarity in recorded 'noise'.

Striking resemblance. Published data from studies of different devices revealed a similarity in recorded "noise". Schön says the bottom figure was sent to *Science* by mistake.

In May 2002, a committee was established to investigate any possible misconduct. When it reported back in autumn of the same year, it had found that Schön had substituted data sets, failed to maintain laboratory records and deleted all of his original raw data. In September 2002 Schön was dismissed from Bell Laboratories.

There is no doubt that Schön was bright and knew his physics. All of his work was peer reviewed and he worked with a large number of co-authors. None of them spotted what was happening, and yet all were cleared of any form of misconduct by the committee that investigated Schön. The process of peer review clearly has its flaws, but it is still hard to know what a better system for evaluating research might be.

Activity 2.1

Allow, about 20 minutes

Read the article 'Conduct unbecoming' below. As you read, think about issues arising from this story that relate to the course themes of *communication* and *ethical issues*. Note down two or three points relating to each of these themes.

Conduct unbecoming

TRAGEDY is the only word that adequately describes the four years that Hendrik Schön spent at Bell Labs in New Jersey. Here was a young man on a roll in one of the world's most prestigious labs. The discovery of a novel way to inject electric charge into organic crystals enabled him to turn insulators into conductors and raise the temperature at which buckyballs superconduct by more than 100°C. He went on to create transistors out of single molecules. Everything he touched turned to gold.

Until last week, that is, when an independent panel employed by Bell Labs concluded that much of his data was fabricated. Reactions from physicists ranged from disbelief to sorrow. Disbelief because Schön's papers promised revolutions in plastic electronics, high-temperature superconductivity and nanotechnology; now no one knows where the frontiers in these fields really lie. Sorrow not only for Schön, whose motives are still unclear, but also for the many researchers who tried to replicate his findings – virtually all without success. Many feel foolish, too, for not spotting Schön's manipulations sooner.

Meanwhile, Schön seems to be taking the fall for his actions alone… Bell Labs sacked him, saying this was the first time in 77 years that anyone there had broken the "scientific honor code". Editors at *Science* and *Nature*, which published 13 of Schön's disputed papers, argue that the events have not exposed flaws in their peer review systems. Bertram Batlogg, Schön's mentor and senior author on many of his papers, has been unavailable for comment since the report's publication.

Yet when rogue operators are exposed, the failings are rarely theirs alone. In the mid-1990s, when Nick Leeson toppled Barings Bank with unauthorised derivatives dealing, the blame spread far and wide. He had been given too much latitude in which to operate. Some of his managers, Barings' supervisory structure and even the Bank of England's system for monitoring banks were all called to account.

So, was Schön also given too much latitude? Many of the ideas for Schön's work came from Batlogg, yet the investigating panel shows that he did not see any of these ground-breaking experiments performed. Only one of Schön's collaborators ever did. What were they thinking? Schön's were spectacular findings, worthy of the toughest scrutiny. Yet there seems to have been a failure of the curiosity and scepticism that are essential to science.

Bell Labs must also shoulder some blame. When Schön met the panel of investigators, he couldn't produce lab notes or computer files showing his raw data. In many labs, it is standard practice for a senior researcher to sign off the lab notes of postdocs, or for experimental results to be stored somewhere that is open to all. This clearly did not happen in Schön's case. Bell Labs announced last week that it would look again at its policies for publishing experimental work and encourage more rigorous internal peer review.

And what of the journals? Donald Kennedy, editor-in-chief of *Science*, reiterated his view that peer review has never been expected to detect scientific fraud. Maybe, but the anomalies in some of Schön's papers to *Science* and *Nature* were hardly subtle: in one, he used the same curve to represent the behaviours of different materials, and in another he presented results that had no errors whatsoever. Both journals stress that papers are chosen on technical merit and reviewers for their technical skills. Should not the manuscript editors or reviewers have remarked on these discrepancies? These papers were, after all, making claims of huge importance to industry and academia.

Ultimately, Schön was unmasked by scientists not engaged in formal peer review. This, some people argue, shows that physics can regulate itself. But it was four years before his actions were discovered, during which some researchers have wasted money and their careers chasing rainbows. The episode has damaged the public's confidence in science – especially physics. Governments and companies will be wary of investing in plastic electronics and nanotechnology. And the conditions that created Schön – the replacement of curiosity with the pressure to publish whenever possible and to find commercial breakthroughs – are spreading everywhere. Physicists need to investigate these issues. If nothing else, it's time for them to follow biologists and create a system for preventing and investigating scientific misconduct.

(*New Scientist*, 2002)

As a footnote to the story of Hendrik Schön, physicists at Rutgers University in New Jersey succeeded in making transistors from organic crystals in early 2005. The work was inspired by Schön's work, and whilst Schön was fraudulent in the way he handled and presented his data, he had good reasons for pursuing organic transistors. A number of respected scientific achievements have had doubtful roots: the reported results of Gregor Mendel's genetics experiments were unfeasibly precise, and the Apollo space programme benefited from technology developed by the Nazis. The process of science is not always what we might like it to be.

2.2 'Smart' and 'functional' materials

Smart materials are those which interact with the environment in a positive way. So, spectacle lenses that darken in bright light are 'smart'. Functional materials are not designed to change with the environment, but are designed to serve specific purposes. Anti-climb paints and anti-scratch coatings are examples of functional materials.

The possibilities that seem to be suggested by working with nanoscale materials have led to some grand claims. K. Eric Drexler has for many years been one of the most forceful but controversial advocates of nanotechnology (more of this in Chapter 4). In an article entitled 'Machine-phase nanotechnology' in 2001, he wrote:

The ability to construct objects with molecular precision will revolutionize manufacturing, permitting materials to be greatly improved. In addition, when a production process maintains control of each atom, there is no reason to dump toxic leftovers into the air or water. Improved manufacturing would also drive down the cost of solar cells and energy storage systems, cutting demand for coal and petroleum, further reducing pollution. Such advances raise hope that those in the developing world will be able to reach First World living standards without causing environmental disaster.

Low-cost, lightweight, extremely strong materials would make transportation far more energy efficient and – finally – make space transportation economical. The old dreams of expanding the biosphere beyond our one vulnerable planet suddenly look feasible once more.

(Drexler, 2001)

In a similar manner, Mark and Daniel Ratner introduce the chapter on 'Smart materials' in their popular science book on nanotechnology, in the following way:

Suppose that corrosion processes could be effectively stopped so that bridges and railroads could be maintained at a fraction of the current cost. Suppose that stain preventers could be incorporated permanently into clothing so that spilled soup would no longer mean a trip to the dry cleaner. Suppose that automobile windshields did not get wet so that no ice would ever form on them and rain could not impede visibility. Suppose that bathroom tiles and hospital sheets could be developed that would self-clean, killing any bacteria or virus that settles on them. Suppose that automobile windows could automatically adjust their reflections to the prevailing sun so that a car parked in a Phoenix afternoon would remain at a civilized temperature. Suppose that a rip in a fabric or a puncture in a tire could immediately and automatically repair itself. All these things are possible, and some are already a reality. They come from the use of smart materials.

(Ratner and Ratner, 2003)

Whilst some of the claims being made for nanomaterials may seem somewhat overblown, there clearly are a number of areas where work is proceeding rapidly and an increasing number of materials-based applications of nanotechnology that are starting to be seen in our everyday lives.

Activity 2.2

Allow 10 minutes

Figure 2.8 illustrates some potential uses of nanotechnology. Make a note next to each numbered item indicating the extent to which you think the application is developed; it may be well developed and perhaps already in use, or it may be a very likely development but still in the early stages, or you may feel it is more in the realm of 'wishful thinking' than likely reality. You might choose to write 'Yes', 'Early' or 'Wishful' by each application. Perhaps as you work through the rest of the chapter you will refine your views on some of these applications.

1. Organic Light Emitting Diodes (OLEDs) for displays
2. Photovoltaic film that converts light into electricity
3. Scratch-proof coated windows that clean themselves with UV
4. Fabrics coated to resist stains and control temperature
5. Intelligent clothing measures pulse and breathing rate
6. Carbon nanotube reinforced frame is light but very strong
7. Hipjoint made from biocompatible materials
8. Nanoparticle paint to prevent corrosion
9. Thermo-chromic glass to regulate light levels
10. Magnetic layers for compact data memory
11. Carbon nanotube fuel cells to power electronics (and vehicles)
12. Nano-engineered cochlear implant

Figure 2.8   Some potential uses of nanotechnology.

2.2.1 Bulk nanostructured materials

A **grain**, or crystal, is the area occupied by a *continuous* crystal lattice. Whilst the whole of a piece of a given crystalline material has the same basic crystal structure, the structure is not continuous throughout the solid; rather, it is formed of a large number of small 'islands', or grains, at the boundaries of which the orientation of the lattice changes. The grain structure of galvanised iron is shown in Figure 2.9. It is worth mentioning that what you see in this figure is the grain structure of the zinc coating that is the 'galvanising'. Smaller grains have proportionally more grain boundaries (this is like the surface area argument you explored in Chapter 1) and boundaries block the type of mobile defects that occur in conventional materials. There isn't scope here for a detailed discussion of defects, but most defects occur when atoms are either missing, or misplaced, from their 'usual' places on the crystal lattice. When such defects occur, other atoms may jump in to occupy the empty space; this leaves an empty space further along the lattice, which is why some defects are mobile: the unoccupied site moves as atoms jump in to fill it. Thus defects can travel through a material until they reach a point where the lattice arrangement changes, a bit like a run in

Figure 2.9   The grain structure of the zinc coating of galvanised iron.

a fabric stopping at a seam. Grain boundaries effectively play the role of seams: they define the point at which the orientation of the lattice changes.

Bulk nanostructured materials are solids that are made up of nanoparticles. These nanoparticles can be disordered with respect to one another, i.e. at random orientations, or they can be ordered in lattice arrays.

There are a number of methods for making disordered nanostructured solids. Nanoparticles are produced and then either consolidated into solids or deposited electrically into sheets. The manufacture of nanoparticles requires specific conditions, often involving high temperatures and carefully controlled environments. Many methods of producing nanostructured solids, therefore, involve large amounts of energy, which inevitably introduces questions of cost effectiveness.

Why are nanostructured solids desirable? What advantages do they offer over materials with conventional grain sizes? In order to understand how nanosized grains affect the properties of materials it is necessary to discuss some of the mechanical properties of conventional materials. The simplest way to investigate the mechanical properties of a solid is to apply a force to a sample to stretch it. There are three distinct ways a solid might respond. It might stretch and return to its former shape when the force is removed; this is called **elastic** behaviour. It might snap, which is known as **brittle** behaviour. Finally, it might stretch past a point where the deformation is reversible and become plastically (irreversibly) deformed; this is known as **ductile** behaviour. All solids show some degree of elasticity in the initial stages of a force being applied.

In the 17th century, Robert Hooke tested a variety of materials by hanging weights on them and measuring the extensions. He observed that for most materials that behaved elastically, the extension was directly proportional to the weight. This observation is now known as **Hooke's law**.

Only in the 19th century was a general method of analysis developed to account for both the size of the sample and the material. If a force, of magnitude F, is applied to the ends of a wire of cross-sectional area A, the wire will stretch. The **stress**, σ, on the wire is defined as:

$$\sigma = \frac{F}{A}$$

so that stress is the force applied per unit area of the wire. The SI unit of stress is the **pascal**, symbol Pa, equivalent to 1 N m^{-2}. The **strain, e,** is the proportional amount the wire stretches as a result of this stress and is given by:

$$e = \frac{\Delta L}{L}$$

where ΔL is the change in length of the wire and L is the original length. The strain is a ratio of two lengths and therefore has no units.

Since Hooke's law states that the relative amount a wire stretches is proportional to the force applied, this means that stress is proportional to strain for elastic deformation. The same is true when a wire, or rod, is compressed along its length within its elastic range:

$$\sigma \propto e$$

The constant of proportionality is called **Young's modulus** and has the symbol E:

$$\sigma = Ee$$

■ Rearrange the above equations to find an expression for E in terms of F, L, ΔL and A.

▨ $E = \dfrac{\sigma}{e} = \dfrac{F/A}{\Delta L/L} = \dfrac{FL}{A\Delta L}$

Young's modulus is a property of a material. You can look up its values for various materials – it does not change unless the material is changed somehow. It is dependent on the separations between atoms that are characteristic of that material. Its value characterises the elasticity of a material. The larger the value of Young's modulus, the less elastic, or stiffer, is the material.

Young's modulus of a nanostructured material is more or less the same as that of the conventional bulk material, until the grain size becomes very small, around 5 nm.

Figure 2.10 shows the ratio of Young's modulus E in nanograined iron to its value in conventional grain-sized iron E_0, as a function of grain size. You can see that as the grain size goes below 20 nm, Young's modulus begins to decrease markedly from its value in conventional materials, which indicates that the material becomes more elastic, i.e. more able to stretch without becoming permanently deformed. This is because there is proportionately so much grain boundary and so little continuous crystal that averaging over the separations between atoms that are typical of the normal crystalline material is no longer valid.

Young's modulus is a measure of stiffness. Strength and stiffness are not the same. Tensile strength is a measure of the amount of stress needed to pull a material apart. Unlike stiffness, tensile strength increases as grain size decreases. Remember that as grain size decreases, there is proportionately more grain boundary and that grain boundaries block the movement of defects. Mobile defects weaken a material and blocking them leads to increased strength.

Bulk nanostructured materials, with which we started this section are quite brittle, as a rule, and not as ductile as their conventional counterparts. Ductility is the ability of a metal to be drawn out into a thin wire, deforming without breaking. You can see then that the variations in mechanical properties that occur by using a nanostructured material make such materials particularly useful for certain applications. Increased strength and elasticity in particular are desirable for many applications and such materials are starting to find use in the automotive and aerospace industries, as well as for sports products and medical implants. Later in this chapter you will look at carbon nanotubes in some details. These unusual nanoscale structures have a Young's modulus as much as ten times greater than steel, making them both strong under tension and very stiff.

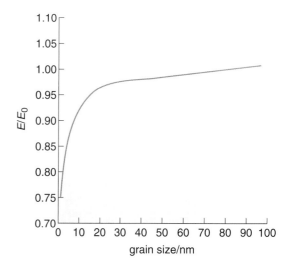

Figure 2.10 The ratio of Young's modulus E in nanograin iron to its value E_0 in conventional iron plotted against grain size. E/E_0 markedly decreases when the grain size is smaller than about 20 nm, indicating that Young's modulus is decreasing, which means that the material is becoming more elastic.

There are other, non-mechanical, properties that change when a material has a bulk nanostructure. There isn't the scope to discuss them all in detail, but here are two examples. Magnetic behaviour changes and this has technological importance related to the possibility of improving magnetic information storage in computer hard drives and other magnetic storage devices. In Chapter 1 you saw how nanoparticles are more reactive than their conventional bulk counterparts, which is again a result of the increased surface area and the number of atoms at the surface of nanoparticles. This increased reactivity is also observed in bulk materials made of nanostructured grains.

2.2.2 Layers and coatings

Very thin films are structures that have one dimension at the nanoscale. In practical terms these are sometimes used as very thin coatings for materials. For example, modern spectacle and camera lenses often have a coating to protect them from skin oils, dirt, dust and scratching.

For current commercial applications most nanolayers are fabricated by chemical and electrochemical deposition methods. The thinnest layers that can be produced are monolayers, i.e. layers a single molecule thick. Whilst not finding much use in current applications, they are a good system for illustrating the concept of self-assembly that you met briefly in Section 1.4.2.

Self-assembly is a bottom-up process whereby atoms and molecules arrange themselves into ordered nanoscale structures. This is not a new idea; crystal growth (including the growth of snowflakes) is an example of a self-assembly process. Self-assembly occurs in many natural examples, and billions of years of evolution have resulted in working molecular machines such as enzymes that put themselves together and then assemble other molecules. One goal for nanoscientists is to design molecular building blocks that spontaneously assemble into desired structures.

The key idea underlying self-assembly is that molecules will always try to occupy the lowest energy level available to them. This is analogous to the way an object rolls downhill (minimising its gravitational potential energy) or a compass needle generally orientates itself in an N–S direction (minimising its energy with respect to the Earth's magnetic field). In manufactured self-assembly, the idea is to exploit this principle so that the energetically preferred bonding or orientation is that which is required, and components naturally organise themselves the way we want them to.

Practical self-assembly relies on the weaker non-covalent intermolecular interactions that you looked at in Section 1.4.2. These weaker interactions allow for self-correction or self-healing which leads to the formation of stable, relatively defect-free, structures. This is because the interactions are often reversible, which makes it possible for defects to be eliminated. The approach hinges on the design of simple building blocks which can incorporate one or more of these non-covalent interactions. The combination of two or more of the non-covalent interactions increases the ability of the building blocks to adapt to different environments and to begin to assemble themselves into ordered structures.

For example, some molecules have **hydrophilic** (water loving) and **hydrophobic** (water hating) portions. When placed on water these molecules, trying to find a state of lowest energy, spontaneously arrange themselves into an ordered monolayer on the water's surface. This is shown in Figure 2.11. You can get a feel for why this is the state of lowest energy by considering the work that would need to be done to turn each molecule around so that the hydrophobic portions were in water and the hydrophilic ones out. Other molecules can do the same sort of trick on different surfaces. A number of different types of self-assembled monolayer have been developed.

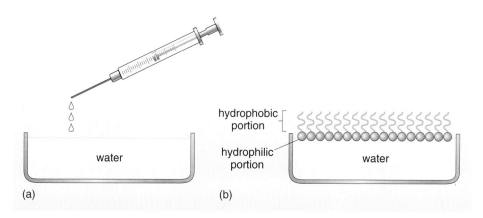

hydrophobic portion

hydrophilic portion

water

water

(a)

(b)

Figure 2.11 Self-assembly of a monolayer (not to scale). To form a molecular monolayer, the material is first dissolved in a low boiling point solvent such as chloroform. This solution is spread as small droplets over the surface of the water using a syringe (a). After a short period, the solvent will have evaporated, leaving a monolayer of material on the surface (b).

Other layers and coatings with thicknesses at the nanoscale, that are not self-assembled, are becoming increasingly widespread. This is one area where 'smart' materials really are finding a place in everyday life. You have already seen, in Section 1.3.4, how nanoparticles become transparent to visible light as their size becomes small compared with the wavelengths of visible light, and this makes them ideal for coating glass and endowing the glass with new properties.

The earliest smart coated types of glass were the low-emissivity (low-e) glasses which have been available since the 1980s. Low-e coatings reduce the heat radiation from window glass and hence from buildings. Low-e glass coatings work by letting sunlight through but not allowing longer wavelength IR radiation to be transmitted back through the glass. This radiation is emitted when the visible light is absorbed within a building by furniture, fixtures and people. These then re-radiate longer wavelength IR radiation, which shows itself as heat. Uncoated glass allows a significant fraction of this radiation to pass through and it is lost, whereas low-e coatings absorb the IR radiation and re-emit the heat back into the room

At the time of writing (2006), the UK Government's target is to reduce CO_2 emissions by 20% by the year 2010 compared with levels in 2000. Pilkington, the glass manufacturers, estimate that replacing all single-glazed windows in the UK with low-e double glazing would represent a saving of 9 million tonnes of CO_2 per year, approximately 7% of the UK's CO_2 reduction target. We don't know what assumptions were made about building use, internal temperatures, air-conditioning and other factors that would affect the outcome of this estimate. Whilst Pilkington may well have a vested interest in this estimation, reducing the heat loss from windows nonetheless seems a worthwhile measure in the attempt to reduce CO_2 emissions.

Another recently developed smart coating for glass is one that will appeal to everyone – self-cleaning glass. Pilkington researchers announced its development in 2001. The coating has several chemical layers to bond it permanently to the glass, but the key component is titanium dioxide (TiO_2). This layer functions in two ways which work together to make the glass self-cleaning. The first effect is photocatalysis. In photocatalysis, what happens is that the TiO_2 (a semiconductor) absorbs energy from sunlight, exciting an electron into its conduction band which leaves a positively charged hole behind. Each hole can migrate to the surface of the TiO_2 layer and participate in oxidation and reduction reactions at the surface. These reactions clean the window by breaking down organic dirt on the glass to form mainly CO_2 and water. Unlike most glass surfaces, TiO_2 is also hydrophilic. This means that it attracts water and causes it to spread on the surface rather than forming separate droplets. As a result any particles of dust or dirt can be washed away from the surface easily. Self-cleaning glass does not clean a surface instantly, therefore; it works continuously using sunlight and rainfall on an ongoing basis.

Before we leave the subject of coatings, it is worth mentioning that the wear- and scratch-resistant coatings used in spectacle lenses, for example, have been significantly improved by nanoscale intermediate layers, or sometimes multilayers. These layers, inserted between the hard outer layer and the substrate material, provide improved bonding and help to match thermal and elastic properties. In the same way that with bulk nanostructured materials the nanostructure blocks the movement of defects that occur in conventional solids, any defects within multilayers are likely to be confined within the layers and this adds to the hardness of the coating.

2.3 Environmental and energy saving applications

In the discussion about smart coatings for glass, you have already seen one example of an environmental application of nanotechnology. With the pressures on businesses to reduce their CO_2 outputs it is easy to see why companies like Pilkington are investing so heavily in technologies like this. Apart from reductions in CO_2, there are a number of other environmental issues driving investment in nanotechnology, not least of which are methods for remediation (cleaning up) of environmental pollution and methods of producing renewable energy.

2.3.1 Environmental remediation

In Section 1.3.1 you saw that increasing the surface area of a heterogeneous catalyst can increase the catalytic activity and also that nanoparticles have a larger surface area relative to their volume than larger particles. This makes them good candidates for efficient catalysts.

The specific surface area of a catalyst, which is a measure of how much area is available for chemical reactions, is usually given in units of $m^2 \, g^{-1}$ and denoted by the symbol S. Note that the units are of surface area per unit mass, not per unit volume. Commercial catalysts, including those made from nanoparticles, have values of S typically in the range $100–400 \, m^2 \, g^{-1}$.

Increasingly, nanoparticles are being incorporated into zeolites. Zeolites are porous materials in which the pores have a regular arrangement. Zeolites have frameworks of silicon, aluminium and oxygen atoms which form channels and cages. It is worth pointing out that zeolites are not new: there are 56 naturally occurring zeolites, found in rocks, and many more synthetic ones. The most common use of zeolites is as ion exchangers (water softeners) for detergents; they are also used as catalysts and to clean wastewaters. An example of the structure of one zeolite, faujasite, is shown in Figure 2.12.

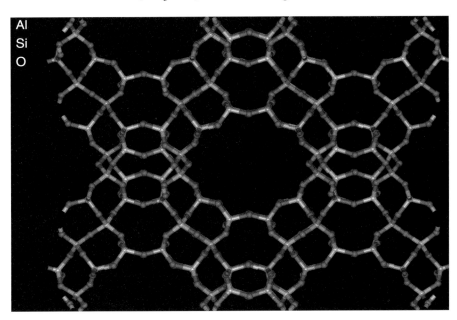

Figure 2.12 A zeolite, faujasite.

The variation in the pore size in zeolites leads to the important application known as molecular sieving. A zeolite is able to accommodate or reject molecules according to their size.

Another application is nanoparticle growth, where the pores of a zeolite are large enough, though within the nanometre range, to accommodate small clusters. This is illustrated in Figure 2.13. The clusters are stable within the pores due to the weak London forces between the cluster and the zeolite. In zeolite catalysis, nanoparticles of a catalyst are found within the pores of a zeolite. This combination can utilise both the high reactivity of the nanoparticles and the molecular sieving effects of the nanopores in the zeolite.

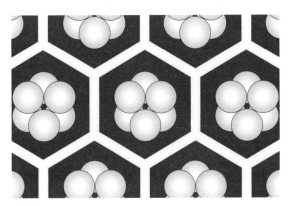

Figure 2.13 Schematic diagram of nanoparticle clusters in zeolite pores.

Wastewater cleaning and other pollution prevention or water-treatment techniques represent a promising, but controversial, aspect of nanotechnology.

There is a good deal of interest in using nanoparticles for purifying drinking water and removing elements such as arsenic. In Topic 3 you learned about the problems of arsenic in groundwater in Bangladesh. In the USA, where the acceptable level for arsenic in drinking water is much lower than in Bangladesh, at 10 ppb, there is commercial pressure to develop technologies that can target and remove such contaminants. A number of different nanoparticles are being used and tested in this regard. Some techniques are simply exploiting the large surface area of reactive materials as catalyst support structures. Others are using the fact that, on the nanoscale, the properties of materials can change. For example, it has been shown that zinc oxide nanoparticles adsorb arsenic whereas the bulk zinc oxide does not. Whilst these applications are very new still, it is clear that there is a good deal of commercial interest in developing them further.

However promising these applications might appear, their use would involve the release of nanoparticles into the environment. In Section 1.3.5 you looked at the risks associated with the inhalation or ingestion of nanoparticles and saw how adherence to the precautionary principle was being advised by the reinsurance company Swiss Re. In fact, the Royal Society and Royal Academy of Engineering 2004 report to the UK Government, was unequivocal on the issue of nanoparticles used in this way:

> Perhaps the greatest potential source of concentrated environmental exposure in the near term comes from the application of nanoparticles to soil or waters for remediation … In some cases the nanoparticles used for remediation are confined in a matrix but, in pilot studies, slurries of iron nanoparticles have been pumped into contaminated groundwater in the USA. Given the many sites contaminated with chemicals and heavy metals, the potential for nanotechnologies to contribute to effective remediation is large. But this potential use also implies a question about eco-toxicity: what impact might the high surface reactivity of nanoparticles that are being exploited for remediation have on plants, animals, microorganisms and ecosystem processes?

This section of the report concludes with the following recommendation:

> We recommend that the use of free (that is, not fixed in a matrix) manufactured nanoparticles in environmental applications such as remediation be prohibited until appropriate research has been undertaken and it can be demonstrated that the potential benefits outweigh the potential risks.
>
> (Royal Society and Royal Academy of Engineering, 2004)

In its response to this report, published in February 2005, the UK Government accepted this recommendation for a precautionary approach and further discussed a review of environmental regulations covering the use of nanotechnologies in environmental remediation.

The question of environmental remediation is complex. The majority of the applications we have looked at in this topic are finding support because they have commercial promise and so are attracting investment. The question of nanotechnology and the developing world hasn't arisen yet. But the question of water purification is of crucial significance in the developing world. In the spring of 2005, a discussion forum, the Global Dialogue on Nanotechnology and the Poor (GDNP), met for the first time. In a briefing paper prepared for the GDNP by the Meridian Institute, an agency that specialises in public policy problem solving, the most compelling and promising aspect of nanotechnology identified is its promise to provide access to safe drinking water. So we have a situation in which new technology seems to have the potential to have a significant humanitarian impact, but where the risks are not understood and caution seems well advised.

■   What are the similarities and differences between this situation and the situation described in Topic 6 with the 'Golden Rice' case study?

▨   It is similar in that it is an application of new technology, which has primarily found commercial interest in the developed world, to a situation where there is the potential for large-scale humanitarian benefit in the developing world. It is different in that, with the Golden Rice case, the problem of vitamin A deficiency can be addressed by increasing the intake of vegetables in the diet such that the 'technological' solution becomes redundant. In the case of water purification, there are no simple, cheap solutions available and the problem is ever more urgent.

The question of how nanotechnology may impact differently in the developed and developing worlds will be picked up again in Chapter 3, with particular focus on health and drug development and delivery.

2.3.2 Renewable energy

Nanoparticles are also finding use in photovoltaic (or solar) cells, which are devices that convert sunlight into electrical energy. Again, some techniques are exploiting the large surface area of nanoparticles. One development uses 1 nm particles of silicon on an amorphous silicon base (which is a commonly used material in photovoltaic cells). In the nanoparticles, electrons can be excited to states in the conduction band of the semiconductor by UV light. Conventional photovoltaics convert only visible light into electrical energy and the UV absorbed is converted just into heat and lost. By using a nanoparticle layer, similar physics to that discussed in relation to quantum dots is exploited, such that electrical energy can be produced from the UV as well as the visible light incident on the nanoparticles.

Another nanoparticle-related development in solar cells is the Grätzel cell, a dye-sensitised solar cell that copies aspects of photosynthesis. In chloroplasts, just as in silicon photovoltaics, the basic principle involved is that photons of sunlight boost electrons into excited states. The problem is that, since the electrons are negatively charged, they leave behind holes, or positively charged equivalents. Left alone these will recombine, re-emitting the absorbed light. If the electrons can be separated from the holes they leave behind, their energy can be utilized. Every solar device must find a way to separate the electrons and holes so that current can be made to flow.

Conventional photovoltaic devices use an electric field to push the opposite charges apart. Chloroplasts, however, use a more subtle approach. They separate charges by distinguishing between the units that generate the electron and those that transport it away. The photosynthetic apparatus inside the chloroplasts is embedded in a membrane. The electron is moved rapidly from one molecule to the next moving it away from where it was generated. It is carried across the membrane before it can recombine with the positive charge. At the other side of the membrane its energy is stored as adenosine triphosphate (ATP).

In the Grätzel cell, organic dye molecules containing ruthenium ions that absorb visible light are coated onto nanoparticles of TiO_2. The nanoparticles do more than just support the dye. The TiO_2 has the right properties to carry the excited electrons away rapidly: it pulls electrons out of the dye and shuttles them off into an electrical circuit. To make a complete solar cell, a 10 μm thickness film of dye-coated nanocrystals is sealed between two transparent electrodes. An **anode** is a positive electrode that attracts electrons and a **cathode** is a negative electrode that repels electrons. The nanoparticles form a colloid, as defined in Box 1.5; this packs a large absorbing surface area into a small space, and is ideal for collecting light. The space between the nanoparticles is filled with a liquid **electrolyte** (a substance that dissociates into free ions in solution to produce an electrically conductive medium) containing iodide ions which provide a source of electrons to replace those knocked out of the dye. This is shown in Figure 2.14

Because nanoparticles are transparent to visible light, as you saw in Section 1.3.4, these Grätzel cells can be developed as coatings for glass – the Swatch watch company is marketing watches that incorporate such cells to power them.

Figure 2.14 The dye-sensitised solar cell, first developed in the early 1990s by Professor Grätzel, has great potential as a cheap and efficient solar photovoltaic device. Titanium dioxide is the material, sensitised to the solar spectrum by ruthenium dye. With such solar cells, white light conversion efficiency of over 10% has been realised. (a) The basic cell composition and (b) the mechanism of light absorption, electron injection and subsequent transport.

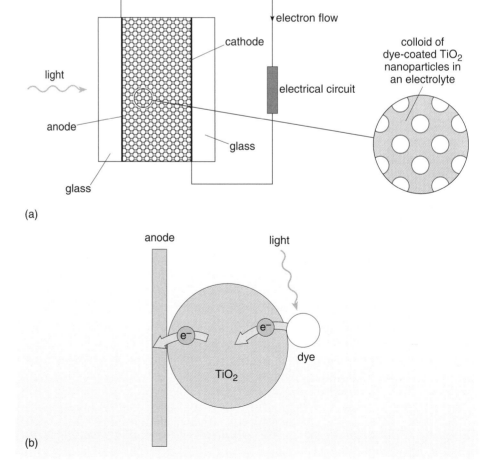

It is not impossible that, with the right dyes and the right advances in technology, this model could lead to the production of windows that also provide electricity for the building.

Nanotechnology is also finding a role in hydrogen fuel cells. A fuel cell is a device that combines hydrogen fuel and oxygen from the air to produce electricity, together with heat and water as by-products. Since the fuel is converted directly to electricity, a fuel cell can operate at much higher efficiencies than internal combustion engines.

A hydrogen fuel cell is shown schematically in Figure 2.15. It is composed of an anode, an electrolyte, and a cathode. When hydrogen reaches the fuel cell anode, the atoms are separated, with the help of a catalyst, into protons (hydrogen ions) and electrons. The electrolyte allows the protons to pass to the cathode side of the fuel cell. The electrons flow to the cathode through an external circuit in the form of electric current. This current can be used to drive an electrical circuit which consumes the power generated by the fuel cell. As oxygen flows into the fuel cell cathode, the oxygen, protons and electrons combine with the help of another catalyst to produce water and heat.

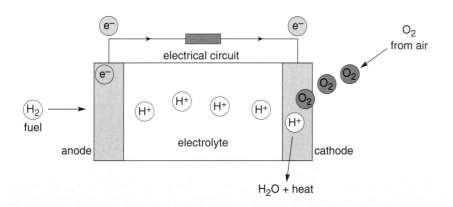

Figure 2.15 Basic fuel cell operation.

The main issue in designing fuel cells is generating and storing the hydrogen fuel. Although hydrogen is the most abundant element in the Universe, on Earth it is mostly found in combination with other elements: in water (H_2O) or fossil fuels like methane (CH_4). So hydrogen must be separated from either water or fossil fuels before it can be used as fuel. The vast majority of hydrogen is produced by so-called 'steam reforming' of natural gas (methane) according to the reaction

$$CH_4 + 2H_2O \rightarrow 4H_2 + CO_2$$

■   What is the obvious disadvantage of this 'steam reforming' method?

▨   The disadvantage is that it produces carbon dioxide as a by-product.

Hydrogen can also be produced from water by electrolysis, a process which uses electrical energy to cause chemical changes. You can see that the problem starts to get complicated. You need some input of energy to create the electricity to produce hydrogen from water in order to use hydrogen as a fuel to produce electricity. The hydrogen production and hydrogen fuel cells need to be efficient to make it worthwhile.

Nanotechnology's potential role in fuel cells is twofold. First, as you have seen, catalysts at the anode side facilitate the separation of hydrogen gas into electrons and protons, and then on the cathode side help with the combination with oxygen to produce water. As you have also seen, nanostructures have the potential to intensify catalytic activity, which could improve the efficiency of fuel cells.

Another possible application of nanotechnology is storing hydrogen in carbon nanotubes. It is estimated that, to be useful, the tubes need to hold 6.5% hydrogen by mass. Such applications are at an early stage of development, but carbon nanotubes and other carbon nanostructures are being actively researched for their potential in hydrogen storage.

There has been scope, in this section, only to look at a small selection of environmental and energy-saving applications. It is inevitable that, over the life of this course, these, and others, will develop rapidly.

2.4 Carbon nanotubes

One day you may look up and see a cable reaching into the sky – something rather like Jack's beanstalk. NASA are thinking seriously about making an earth-to-orbit elevator: a cable extending into space along which electrically powered vehicles would travel. It might even be used to tether a satellite.

Of the myriad materials currently available to NASA engineers, only one is suitable for this task. And it comes as tubes that are 50,000 times thinner than a human hair. The slender proportions of the carbon nanotube hide a staggering strength: along its length, a nanotube is as strong as a diamond.

(Brooks, 2000)

In Chapter 1 you learned about the discovery in 1985 of carbon fullerenes and in 1991 of carbon nanotubes. Over the next 10 years the pace of research into carbon nanotubes was astounding, and we now know that they possess the following properties. They are:

- extremely strong
- able to conduct electricity or not according to their design
- extremely rigid
- capable of storing and enclosing atoms and molecules of other materials
- unusually good thermal conductors
- possibly electrical superconductors
- ideal components in electrical circuitry.

A superconductor is a material that can conduct electricity without having any resistance, so that there is no loss of energy within the material as heat. You are probably familiar with the fact that most electrical conductors that we commonly use have some resistance and get hot. If this resistance could be eliminated, then a current could keep flowing indefinitely without loss. This would make circuits easier to design as there would not be the need for cooling, and they would also consume far less energy.

You might find it useful to look back at Box 1.1 and Section 1.4.1 and make sure that you are familiar with the structures of diamond, graphite, C_{60} and carbon nanotubes.

2.4.1 The structure of carbon nanotubes

In Chapter 1 you read that carbon nanotubes consist of a layer of graphite (which is also called a graphene sheet) rolled up to form a seamless tube with the ends of the tube capped by a hemispherical 'half buckyball'. However, it turns out that the unusual and variable properties of nanotubes depend quite sensitively on the way in which the graphene sheet is rolled. Before looking in detail at the various types of nanotube it is important to be familiar with some of the basics of **vectors**. These are outlined in Box 2.3. You will then use the vector ideas to work out, for yourself, the essential differences in structure between the different types of nanotube. Note that a vector in the physical sciences has quite a different meaning to that used in the life sciences. In Topic 6 the term vector was applied, for example, to a plasmid that transfers genetic material into a cell, whereas here it describes a quantity defined by both its magnitude and direction.

Box 2.3 Introducing vectors

Although vectors may be new to you in terms of mathematics, you will have used them every time you have described the position of something. For example, if you say that Edinburgh is 75 km east of Glasgow you are defining a vector that joins Glasgow to Edinburgh and has a length of 75 km and a direction E. So, to describe positions in two dimensions we need two parameters, the length or magnitude, and the direction. Quite a few quantities you are familiar with need to be defined both by magnitude and direction – displacement, velocity, acceleration and force are all examples of vector quantities. Quantities that are defined only by their magnitude, or size, but have no direction include speed, mass and energy. Such quantities are called **scalars**.

We can represent a vector by a straight line, where the length of the line reflects the magnitude of the vector and its direction is indicated. Vectors can be denoted by bold type or by labelling the two end points, for example A and B, so the vector would be written as $\boldsymbol{a} = \overrightarrow{AB}$. All vectors with the same length and direction are the same vector so, on Figure 2.16, $\boldsymbol{a} = \overrightarrow{AB} = \overrightarrow{CD}$.

When writing by hand, vectors are denoted by underlining, so $\boldsymbol{a}$ is written a̲.

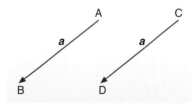

Figure 2.16 All vectors with the same length and direction are the same vector.

We can multiply a vector by a scalar, and this changes its length but not its direction, e.g. as shown in Figure 2.17.

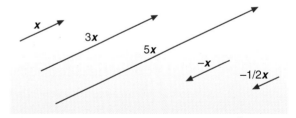

Figure 2.17 Multiplying a vector by a scalar changes its length but not its direction.

Notice that when you multiply a vector by −1, the resultant vector has the same magnitude but the opposite direction, so $\boldsymbol{x}$ and $-\boldsymbol{x}$ have the same length but point in opposite directions (Figure 2.17).

Another useful idea is that of the unit vector. A vector of unit length in the direction of the vector a is called the unit vector and is denoted by $\hat{a}$.

If we go from P to Q by travelling 10 km NE and then from Q to R by travelling 5 km W (Figure 2.18), we can find out the position of R relative to P by drawing the two vectors $\overrightarrow{PQ}$ and $\overrightarrow{QR}$ to scale. Then we can construct the vector $\overrightarrow{PR}$, which is the vector sum of the two previous vectors. We can write this as

$$\overrightarrow{PR} = \overrightarrow{PQ} + \overrightarrow{QR} \quad \text{or} \quad r = p + q$$

In a similar way, vectors can be subtracted as in $r = p - q$. In order to draw this, rewrite this equation as $r = p + (-q)$ and then add the vector $-q$ to p.

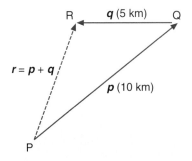

Figure 2.18 The vector $\overrightarrow{PR}$ is the sum of the vectors $\overrightarrow{PQ}$ and $\overrightarrow{QR}$.

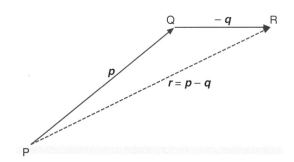

Figure 2.19 Vectors can be subtracted as well as added.

Question 2.2

Dr Vector's route to work involves him driving 4.9 km west, then 3.9 km south, and then 1.8 km east. Draw a scale diagram and determine the magnitude and direction of his resultant displacement.

The structure of nanotubes is most easily visualised in terms of the unit cell of graphite – the smallest group of atoms that defines the structure. Using Figure 2.20 you should be able to see that each carbon atom has three nearest neighbour atoms. Each carbon atom either has one neighbour above and to the right, another below and to the right and one directly to the left as shown in Configuration A in the margin or has one neighbour above and to the left, one below and to the left and one directly to the right as shown by Configuration B in the margin. Either of these configurations can be used to define the unit cell. You can see that each of the labelled atoms on Figure 2.20b 'sees' the same configuration of neighbour atoms. You can also see two vectors, marked $\hat{a}_1$ and $\hat{a}_2$ on Figure 2.20. $\hat{a}_1$ and $\hat{a}_2$ are unit vectors, and they define the directions needed to discuss the hexagonal lattice structure of the graphene sheet. Combinations of $\hat{a}_1$ and $\hat{a}_2$ allow us to define vectors around the lattice. You can see in Figure 2.20b that each marked atom has two numbers in brackets below it. These represent, respectively, the number n of unit vectors $\hat{a}_1$ and the number m of unit vectors $\hat{a}_2$ that the labelled atom is away from the origin. In this system, the origin is therefore labelled (0, 0). Trace for yourself $5 \times \hat{a}_1$ and $3 \times \hat{a}_2$ and check that you do indeed arrive at the atom labelled (5, 3).

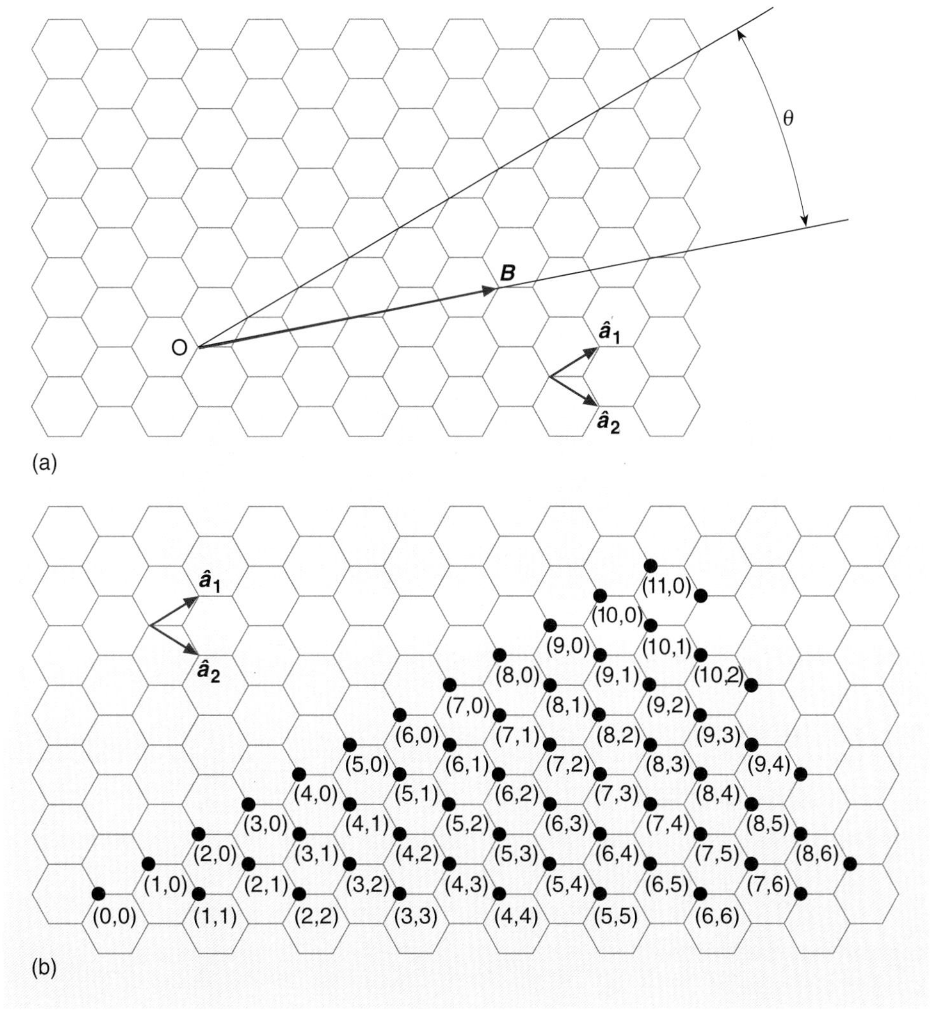

Figure 2.20 Basics of structure of carbon nanotubes.

If you draw a line of any length in any direction (vector $B = n\hat{a}_1 + m\hat{a}_2$) on the graphene sheet, you can join the ends of that line (vector B) together by rolling up the sheet. The direction of the vector determines how much twist, if any, is in the sheet and the length of the vector determines how tightly rolled up it is. You might find it helpful to have one of the transparencies marked with the graphene structure to hand at this point just to help illustrate the idea. Since vector B effectively determines the amount of twist, or **chirality**, of the tube it is called the **chiral vector**. One final definition, the **chiral angle**, θ, is the angle between the chiral vector and the unit vector $\hat{a}_1$, as shown in Figure 2.20a.

Question 2.3

It is important that you do this question and check your answer before embarking on Activity 2.3.

Use one of the blank graphene sheets that you have and choose an origin, O, near the bottom left-hand corner to work out the values of n and m that have been used to construct the chiral vector B in Figure 2.20a.

Activity 2.3

Allow about 45 minutes

In this activity you are going to explore the different structures of so-called armchair, zigzag and chiral nanotubes. You will need to use the printed transparencies that you have; there are three of these that simply have the 'blank' graphene structure printed on them. Use the sheets in 'landscape' orientation. You will also need a ruler, water-soluble transparency pen and some tape.

(a) Starting at the bottom left, draw the chiral vector A (for armchair) such that m and n are both 5. It doesn't matter whether you do all the $\hat{a}_1$ first and then all the $\hat{a}_2$, or whether you do them alternately until you have done five of each. What is the direction of the vector? Roll up the sheet so that the ends of the vector touch. You have made an armchair nanotube which should look like Figure 2.21a.

What would happen if n and m remained equal, but were both equal to, say, 2, or 12, or 20?

(b) Now make $n = 10$ and $m = 0$ and draw the resulting vector Z (for zigzag). What is the direction of this vector? Roll the sheet up so that the ends of the vector meet. This is a zigzag nanotube and should look like Figure 2.21b.

(c) Make $n = 0$ and $m = 10$. Is this different from the tube you made in (b)?

(d) Now try different values of n and m. Call the vectors you draw C (for chiral). You should find that all the chiral vectors you can make lie either on the armchair or the zigzag structure or somewhere in between. Your 'in between' tubes should look something like that in Figure 2.21c. What is the angle between the chiral vector of the armchair tube and that of the zigzag tube?

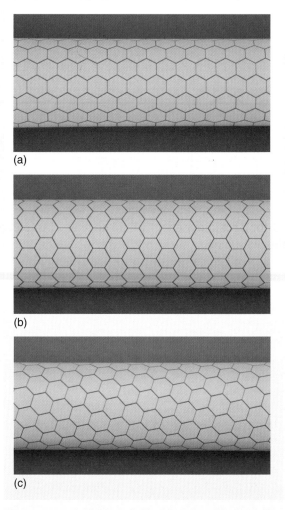

(a)

(b)

(c)

Figure 2.21 Models of nanotubes: (a) armchair structure, (b) zigzag structure and (c) chiral structure.

There are lots of possible chiral structures, depending on the choice of n and m. However, there is only one of each of the zigzag and armchair structures, lying at each end of the possible range of chiral angles.

The properties of nanotubes depend on two parameters: the chiral angle, which you should be starting to have a feel for by now, and the diameter. That the diameter should affect the properties of a tube may seem quite natural, but that the chiral angle – which defines the chirality (or amount of twist) the tube has – should affect its properties is much less obvious and is one of the things that makes nanotubes so unusual and interesting. In fact, the diameter depends on the chiral angle too: you may have noticed in the activity that the chiral angle associated with the armchair configuration takes the shortest path across the graphene sheet, which leads to a smaller diameter. This length increases to the maximum in the zigzag configuration. Sadly, making real nanotubes with chosen diameters and chiralities is still a dream. When nanotubes are made, a random selection is produced. Whilst we can understand how nanotube properties vary with these parameters, we cannot yet manufacture them at will.

2.4.2 The properties of nanotubes

Electrical properties

The most interesting property of nanotubes is that they can be either metallic or semiconducting, depending on their diameter and chirality. Their unique electronic properties are due to the confinement of electrons within the cross-section of the tube. When an electron is within a normal flat graphite sheet it can move in two dimensions. Once the sheet is rolled it is able to move only in one dimension, along the axis (length) of the nanotube. You might find it helpful to look back at Section 2.1.3 for a reminder about what happens when electrons experience confinement at the nanoscale.

As with quantum dots, where the number and spacing of available energy levels depend very sensitively on the size of the semiconductor crystal, so here the number and spacing of energy levels depend on the circumference of the nanotube. The circumference depends, of course, on the diameter, and the diameter, as you have just seen, depends on the chiral vector. The calculation is beyond the scope of this course, but it turns out that, in general, an (n, m) carbon nanotube will be metallic, and so will conduct electricity, only when $n - m = 3q$, where q is any integer, including zero.

■ Using the relationship $n - m = 3q$, what proportion of armchair nanotubes are metallic?

▨ All of them. Since $n = m$ for all armchair nanotubes, and rearranging this gives $n - m = 0$, this will satisfy $n - m = 3q$ with $q = 0$ for all values of n and m.

■ What proportion of zigzag and chiral tubes are metallic?

▨ One-third. Since n and m are integers, $n - m$ can take any integer value. Every third integer value, according to the equation above will lead to a metallic nanotube, so where $n - m = 0, 3, 6, 9$, etc. the nanotube is metallic. This means that one-third are metallic.

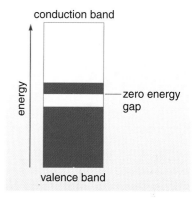

The two-dimensional graphene sheet has a very unusual electronic structure. It is a semiconductor with a zero band gap. This means that the top of the valence band and the bottom of the conduction band have the same energy. However, in the case of graphite there are very few allowed transitions that enable electrons to move into the conduction band. Such a material is known as a **semimetal**. The band structure of a semimetal is shown in Figure 2.22. However, when the graphene sheet is rolled into a tube, this changes the conditions because the electrons are confined to move in only one direction and the graphene can become metallic or semiconducting. For those tubes that are semiconducting, the size of the band gap is inversely proportional to the tube diameter and approaches zero as the diameter gets larger.

Figure 2.22 Band structure of a semimetal. The top of the valence band and the bottom of the conduction band have the same energy, but there are few allowed transitions enabling electrons to move into the conduction band.

In the metallic state, the electrical conductivity of nanotubes is very high. It is estimated that they can carry a current of 1×10^9 A cm^{-2}. In conventional conductors, resistive heating occurs due to the defects in the metal lattice that you met briefly in Section 2.2.1. In Chapter 1 you read about the electron gas as a model for a metal. The ability of electrons to move easily through a metal in this model is due to the fact that the metal lattice is basically uniform. However, where there is a defect, such as an extra or missing atom, electrons see this defect as an obstacle and are scattered from it and the scattering causes a friction-like heating. As the metal gets hotter, the atoms within it also vibrate more and this leads to more collisions, increasing the scattering still further. Copper wire fails at 1×10^6 A cm^{-2} because the resulting heating melts the wire. But carbon nanotubes have very few defects to scatter electrons, and so very little heating occurs. This means that high currents do not heat the nanotubes as they do copper wires. Nanotubes are also very good conductors of heat.

Mechanical properties

Carbon nanotubes are very strong. In Section 2.2.1 you met the concept of Young's modulus. Remember that the larger the value of Young's modulus, the stiffer the material is. The Young's modulus of carbon nanotubes is nearly 10 times that of steel. This implies that carbon nanotubes are very stiff and very hard to bend.

■ Suggest a reason why the above statement might not be true.

▨ The diameters of nanotubes are extremely small compared to their length and so they can in fact be bent relatively easily.

When carbon nanotubes are bent they are very resilient and do not break. They buckle like straws but can be straightened back without damage. This is largely because they have very few defects in their structure and the carbon–carbon bonds are able to withstand the bending without breaking.

As well as having a high Young's modulus, carbon nanotubes also have a high tensile strength. Remember that this is a measure of how hard it is to pull something apart.

■ From what you learned earlier in Section 2.2.1, why do you think carbon nanotubes have such a high tensile strength?

▨ Tensile strength is decreased when defects can move through a material. Carbon nanotubes have very few defects so there is very little defect mobility, increasing tensile strength.

The actual value of the tensile strength varies according to whether the carbon nanotubes are single- or multi-walled (see Section 1.4.1), but carbon nanotubes do have the highest tensile strength of any material yet measured, with the highest single measurement of the tensile strength of a nanotube being 63 GN m^{-2}. This is about 30 times greater than that of high-strength steel. By 2004, however, no object had been constructed using a nanotube-based material that has had a tensile strength anywhere near this value and the improvements to materials from the use of nanotubes are still relatively modest.

2.4.3 Applications of carbon nanotubes

Almost every week sees publication of new developments in the use of carbon nanotubes. This section covers just a few key areas where, at the time of writing (2006), research is very focused.

Field emission

When a small electric field is applied along the length of a tube, electrons can be emitted at a high rate from the ends of the tube. This is known as *field emission*. One application of this effect with carbon nanotubes is the development of flat screen display panels. In such displays, individual nanotubes provide electrons to illuminate a single pixel in the display. The resulting displays have high brightness while being low-energy and lightweight. Another is the development of vacuum tube lamps which exploit this effect and are as bright as conventional lightbulbs but longer-lived and more efficient.

Computing

There is tremendous interest in the use of carbon nanotubes in various aspects of computing. Almost all of them are too complex for more than a passing mention here, but the fact that nanotubes can be such good conductors, together with their small size, makes them ideal candidates for interconnecting wires and small switches. With ordinary metals, as the cross-sectional area of a wire decreases, heat generation becomes a big issue. Carbon nanotubes with diameters of 2 nm have extremely low resistance, which makes them very desirable as connections. Since they also conduct heat very well they can be used as heat sinks to carry heat rapidly away from a computer chip.

Other ideas include making computer components out of carbon nanotubes – some of which would be semiconducting and some metallic. For example, a nanotube formed by joining together two nanotubes with different diameters can act as a diode. There isn't the scope here to go into more detail, but you can start to see how the electrical, mechanical and thermal properties of nanotubes make them very interesting in computer design and manufacture, where all the pressure is to go smaller.

Fuel cells

You saw in Section 2.3.2 that storing hydrogen in nanotubes is another possible application. There is real commercial interest in the development of fuel cells as sources of electrical energy for future cars, computers and mobile phones. Fuel cells need a source of hydrogen, and one possibility is to store this inside carbon nanotubes.

Mechanical reinforcement

Given the high tensile strength and large length-to-diameter ratios of carbon nanotubes, they are excellent candidates for the reinforcement of composite materials. Early work in this area is also promising. For example, a study at the University of Tokyo showed that incorporating 5% by volume of nanotubes in aluminium increased the tensile strength by a factor of two compared with pure aluminium. Theoretical estimates suggest that incorporating a 10% volume of nanotubes should increase tensile strength sixfold if fabrication techniques can be optimised. Other promising test results indicate that the tensile strength of steel could be increased by seven times if the steel had 30% by mass of oriented nanotubes incorporated in it. It is early days still for all of these applications. The expense in making large quantities of nanotubes is a major hurdle to be overcome before we are likely to see real, rapid development of materials incorporating them.

Biotechnology

Although biological applications of nanotechnology will be discussed in detail in Chapter 3, any overview of carbon nanotubes needs to mention that they can be opened and filled with materials such as biological molecules, which makes them candidates for use in biotechnology applications.

This look at the likely development of applications involving carbon nanotubes has been rather brief and cursory. It is a fast-moving area that is bound to change substantially over the life of this course. But we have tried to flag up areas where development is likely to be most interesting and we have given enough of the background science so that you should be able to understand such developments as they evolve.

Summary of Chapter 2

Nanotechnology of materials is being driven by a number of factors. One of the most pressing is the need to find a new approach to miniaturisation in the electronics industry. Nanotechnology offers a fundamentally different approach to the silicon-based electronics we currently have. Quantum confinement of electrons at the nanoscale presents the possibility of using quantum dots which have 'tunable' energy levels. Quantum wires, for similar reasons, present the possibility of low-resistance, high thermal conductivity connections needed at this scale. This research is developing fast, but there are few applications finding their way into active use.

Smart materials are with us already to some extent. Composites have been incorporating nanoparticles for some time, and layers and coatings for glass are developing rapidly and are increasingly common. Biocompatible materials are also finding use. Most developments in this area reflect how nanotechnology is improving existing technologies.

Energy efficiency and environmental remediation is another major driver for nanotechnology and materials. Progress here has a good deal of promise, but in the area of remediation there is concern about the risk of releasing free nanoparticles into the environment and a precautionary approach is being advised by the UK Government.

Carbon nanotubes are a relatively new material, and their properties and potential uses are of great interest to researchers. Their ability to be made metallic or semiconducting, and their great strength and other properties mean that they are being investigated for use in a huge range of applications. We can expect this field to continue to expand rapidly.

Questions for Chapter 2

Question 2.4

Explain why the electrical conductivity of a metal decreases when the temperature increases and why the electrical conductivity of a pure semiconductor increases with increasing temperature.

Question 2.5

A vertical circular steel post has a diameter of 15 cm and is 3.0 m long. It supports a load of 6.0×10^3 kg. Use the value of 2.0×10^{11} Pa for the Young's modulus of steel and the value $g = 9.8$ m s^{-2} for the acceleration due to gravity to calculate:

(a) the stress in the post;

(b) the strain in the post;

(c) the change in the post's length.

Question 2.6

The Royal Society and Royal Academy of Engineering report to the UK Government recommends that the use of free manufactured nanoparticles in environmental applications be prohibited until it has been demonstrated that the potential benefits outweigh the potential risks. What are the two main reasons for this caution?

Chapter 3

Nanotechnology and living systems

In previous chapters, nanotechnology has been portrayed largely as the province of the physical sciences. But, as this chapter will make clear, the techniques and approaches you have read about are being increasingly applied to the biological sciences, with new links being forged between the fields of life sciences, genetic engineering (of the type you studied in Topic 6) and information technology. Most probably, it is in this newly emerging territory that many of the most radical innovations will become reality within a time span as short as 20 years or so. But predicting any more precisely what areas of 'biological nanotechnology' will prove especially productive, and on what time scale, would be a risky undertaking, given that research and development is proceeding across such a wide range of different areas.

In this chapter we will look at a range of examples of the application of nanotechnology to living systems, saying a little about each. The intention is to look at a good number of examples, though none in any great detail. This means that the style of what follows is different from that of Chapters 1 and 2. We feel that this more wide-ranging (and less quantitative) approach is better suited for a topic where there are so many different applications awaiting development. To help you identify the key points within the broad areas covered, the early sections of the chapter each end with a summary paragraph. We will be looking primarily at examples of likely medical benefit, especially later in the chapter, and explaining why the prospect of changing human lives so dramatically in the not too distant future raises a mix of excitement and concern. As we will see, living systems contain a great wealth of 'machines' that function at the nanoscale, of a type and variety that evoke inspiration and awe in those enthusiasts currently struggling to manufacture artificial counterparts.

3.1 The living cell as a nanotechnology factory

Much of cell biology functions on the nanoscale. If you refer back to Figure 1.1, you will appreciate that many biological substances fall within the nanoparticle range of between 1 and 100 nm. Some whole organisms are close to that range; the largest bacteria (at about 0.75 mm) are at the upper end of the microscopic scale but the smallest (at about 20 nm) are genuine biological nanoparticles. The same term can be applied to many viruses, with dimensions that vary from 10 to 200 nm.

Table 3.1 provides more precise information on a range of biological substances. Individual amino acids are below the nanoscale size range (at about 0.6 nm), but proteins can be thought of as nanoparticles. Given the enormous variety of functions they display, the cellular structures that produce proteins must themselves be very impressive manufacturing nanomachines. DNA is perhaps the biological nanoparticle *par excellence*. Indeed, some talk of DNA as two nucleotide 'nanowires', twisted around each other, with a diameter of 2 nm. Incidentally, the parallels are stronger than those of simple nomenclature alone.

There is interest in the ability of DNA to transport electrons along the length of a few base pairs, and hence in using the macromolecule as a conducting wire, opening up the possibility of using DNA in computing, perhaps even utilising DNA's powers of self-replication.

Table 3.1 Sizes of a variety of biological substances. In each case, size d refers to the longest dimension. Note that some proteins are rounded, e.g. lipoprotein, with a diameter of approximately 20 nm, while others are flattened and elongated, such as fibrinogen, which is over 80 nm in length

Class	Material	Size d/nm
amino acids	glycine (smallest amino acid)	0.42
	tryptophan (largest amino acid)	0.67
nucleotides	cytosine monophosphate (smallest DNA nucleotide)	0.81
	guanine monophosphate (largest DNA nucleotide)	0.86
	ATP (energy source)	0.95
other molecules	stearic acid $C_{17}H_{35}COOH$	0.87
	chlorophyll	1.1
proteins	insulin, polypeptide hormone	2.2
	haemoglobin, carries oxygen	7.0
	albumin, in white of egg	9.0
	elastin, cell-supporting material	5.0
	fibrinogen, for blood clotting	80
	lipoprotein, carrier of cholesterol	20
	ribosome (where protein synthesis occurs)	30
	glycogen granules (an energy reserve mainly within the liver)	150
viruses	influenza	60
	tobacco mosaic	120
	bacteriophage T_2	140

What is of great interest, therefore, is the way in which these biological nanoparticles are assembled inside the living cell. Very small building blocks (sometimes individual molecules) come together to make particular structures. Such processes epitomise the type of 'bottom-up' nanotechnology that you will know about from Chapters 1 and 2.

Question 3.1

Think of a couple of examples of biological nanomachines that you are familiar with already. You might choose to look back at any notes you made on the movie material viewed at the end of Chapter 1. Section 2.1 of Topic 6 would also be a helpful source of examples.

You know from Topic 6 that proteins play a central role in living systems and that a great many different types are apparent. Some are involved in communication between cells (e.g. receptor proteins), some sit within **biological membranes**,

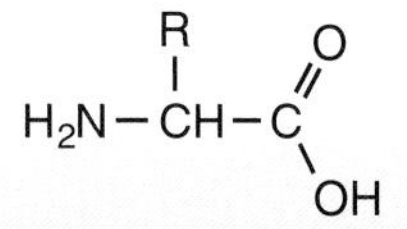

Figure 3.1   Structure of a generalised amino acid.

e.g. transporting ions, while as many as 10 000 different types of enzyme occur within the average cell. Yet there are only 20 amino acids that comprise the individual components from which proteins are manufactured within the cell. Their detailed structure need not be of concern, but it is helpful to have in mind the general formula of an amino acid, as shown in Figure 3.1. Different amino acids vary in the side chains they contain, represented by the R group.

The fact that so many distinct types of protein can be produced, from the same relatively few starting materials, shows the enormous manufacturing potential of the ribosome. This is just one reason why nanotechnologists are inspired by examples from the biological world. Furthermore, as was mentioned in Chapter 2, the nanoscale machines that operate within living cells have evolved over many, many millions of years. Natural selection has produced solutions to the particular operational problems evident at this scale – solutions that can perhaps be mimicked in analogous systems produced artificially. An approach of this type, which builds upon what is known of biological nanomachines, is generally termed 'soft' design, as opposed to the 'hard' engineering design philosophy epitomised by the ideas of Eric Drexler and his supporters, introduced in Section 2.2 and discussed further in Chapter 4. Many nanotechnologists, including the increasing numbers from a biological background, seem attracted to the field of soft design by virtue of the intellectual challenge of outdesigning evolution.

■ To what extent might biological systems be considered 'perfect' and, therefore, superior to artificially manufactured structures?

▨ Biologists are very dismissive of the notion of biological perfection. Efficient and diverse though biological nanoscale machines are, there is no certainty that their abilities cannot be improved upon. (One nanotechnologist commented that 'biology never invented the digital computer, internal combustion engine, or even the wheel'!)

One further aspect of biological nanoparticles that increases their appeal is their capacity for self-assembly, which you know of from Chapters 1 and 2. When the term relates to biological systems it usually refers to a capacity to assemble complex structures autonomously, i.e. without significant external intervention. In its strictest interpretation, only those constituents of the final structure take part in self-assembly. Remarkably, some simple viruses (biological nanomachines towards the upper end of our nanometre range, see Table 3.1) can be constructed via self-assembly alone, from diverse biological macromolecules such as proteins, DNA and RNA.

■ Recall in broad terms the process of DNA replication. Is this an example of self-assembly in living systems?

▨ Indeed it is, although enzymes, notably DNA polymerase, are required to promote the process (Section 2.1 of Topic 6). The separation of the two strands leads to unmatched bases becoming exposed, at which point 'matching' base pairs become positioned, to reconstitute two identical 'daughter' strands of DNA.

But the properties of proteins provide especially intriguing forms of self-assembly, as revealed by Figure 3.2. You will know that each type of protein comprises a unique and precise linear sequence of amino acids (see Figure 3.2a), generally referred to as a polypeptide chain. That sequence will determine the characteristic three-dimensional shape of the resultant protein, acquired during a process of folding that is a remarkable, but as yet poorly understood, form of self-assembly. So, primary structure (Figure 3.2a) determines higher-order structure (Figure 3.2b–d), which means that proteins of a particular composition will generally adopt an exact and consistent three-dimensional structure. As an example, Section 3.1 of Topic 6 described the higher-order structure of insulin, where you might recall that the final product comprises two **polypeptides**, the A and B chains, joined by two disulfide links.

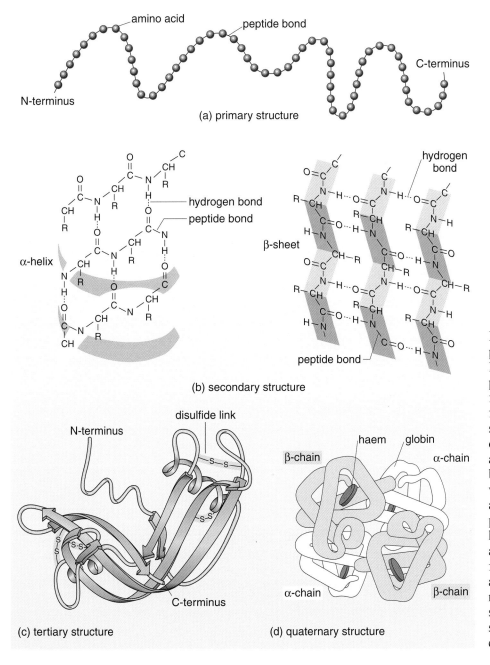

Figure 3.2 The structure of proteins can be defined on four levels. (a) Primary structure, a particular sequence of amino acids forming a polypeptide. Different forms of folding produce a secondary structure, two forms of which are shown in (b), one an alpha (α) helix and the other a flat beta (β) sheet. The amino acids will have distinct R groups; adjacent amino acids are linked by peptide bonds, as shown. The higher-order structures (tertiary and quaternary) arise from further folding, often with disulfide links, as in the protein bovine pancreatic ribonuclease (RNAase) (c), and sometimes the association of several polypeptides or other components, as in haemoglobin (d).

Proteins do precisely the job they do because of their characteristic shape. You will recall from Topic 1 that BSE (and its human counterpart vCJD) both come about because of alterations in the three-dimensional shape of proteins, forming insoluble clumps that seem to stick together. More generally, if one or more amino acids in the sequence of an enzyme are altered, particularly those close to the site at which the enzyme binds to the substrate (i.e. the **active site**), then the enzyme may no longer function, generally because of a change in the three-dimensional shape. If an enzyme is moderately heated, or if the pH of the medium is altered, then the activity of the enzyme is likely to change, largely because of the resultant change of shape. Chemicals that reduce the potency of enzymes (more technically, denature them) will alter the three-dimensional shape, with a resultant loss of activity; remove such a chemical, and the original protein shape can generally be restored. The key point is that the three-dimensional structure of a protein is generally an 'inbuilt' consequence of that particular sequence of amino acids and the interactions between individual amino acids that result.

This capacity of a chain of amino acids to self-assemble into a defined shape by characteristic folding is just beginning to be investigated by physicists and chemists. In the long term, the hope is that proteins of particular shapes could be artificially constructed, in a controlled and predetermined way, to perform a particular nanoscale task – perhaps within living cells. One attractive option is creating new molecular machines in the form of enzymes to undertake new cellular tasks, created by hijacking the normal working of ribosomes to manufacture novel proteins. Proteins produced in such a way might include amino acids other than the 20 that normally contribute to protein manufacture within cells. Indeed, 'protein engineers' are keen to point to the huge array of non-natural amino acids, not found in living systems, but which can be synthesized by chemical methods. The notion of artificial 'purpose-built' proteins is many years away from fruition, but this is not because the notion is intrinsically far-fetched; rather, it is more that very complex physics and chemistry have to be understood before such a synthetic biological nanomachine becomes a reality.

Indeed, it is already possible to apply the process to living cells, at least bacterial cells, which can be supplied with manufactured transfer RNAs and an appropriate enzyme and thereby manufacture proteins with one or more novel amino acids (other than the 20 that occur within cells). This may seem a modest modification, but there is no doubting the potential of the technique; one leading author (Goodsell, 2004) describes this as '…the first steps towards making direct changes in the basic processes in life, creating entirely new organisms different from anything found in the natural world'.

Think of a modest-sized protein, 100 amino acids long. For the 20 naturally occurring amino acids, it is possible to calculate the number of different sequences that are possible for assembly into a protein of this length; the answer is 20^{100}. This is an extraordinarily large number. The author Richard Jones points out that if a single protein molecule were constructed for each sequence, such a collection of proteins would exceed the total mass of the Universe. If novel amino acids were included, the number of possible sequences would be unimaginably greater. However, the most useful proteins are those that fold into a precise and stable three-dimensional structure. Although the chances of any one such stable protein proving of practical value are extraordinarily small, computer simulation is now being used routinely to identify 'useful' amino acid sequences, techniques that should ultimately yield novel nanomachines.

As is evident from Figure 3.2d, highly folded proteins themselves have capacity for self-assembly; many can readily form associations with identical or different proteins, increasing the size and perhaps the complexity of the biological machine that results. Ribosomes are a good example; each consists of more than 50 distinctive types of protein molecule, along with several different types of RNA. Each of the constituent protein types has a highly folded shape that is determined by the particular amino acid sequence; some exposed parts of the protein have 'sticky patches' that are areas where neighbouring proteins become loosely attached.

Proteins sometimes associate with lipids, which you will know as the fatty components of living cells. Indeed, lipids themselves have the capacity for self-assembly, in that they aggregate together to form a sheet-like bilayer, foreshadowing the structure of the more complex biological membranes that you know something of already from Topic 6. Such lipid bilayers have carbon-rich hydrophobic components tucked inside, well away from contact with water and the hydrophilic heads exposed on each of the two surfaces of the sheet. You came across the terms hydrophobic and hydrophilic in Section 2.2.2, in the context of simple monolayers.

Lipoprotein complexes are key components in the biological membrane that surround cells, i.e. the cell membrane, as is apparent from Figure 3.3. Its structure is more complex than the simple bilayer just described, in part because the major fat constituents are phospholipids, so named because they contain a phosphate group. Proteins are a significant component, many spanning the phospholipid bilayer in which hydrophilic heads make up the membrane's inner and outer surface.

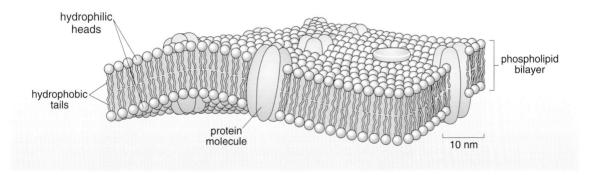

Figure 3.3   A biological membrane, showing the contained membrane proteins (in blue) and the arrangement of the phospholipid components. Some of the membrane proteins span the membrane, while others sit on the surface. Biological membranes are typically about 10 nm wide.

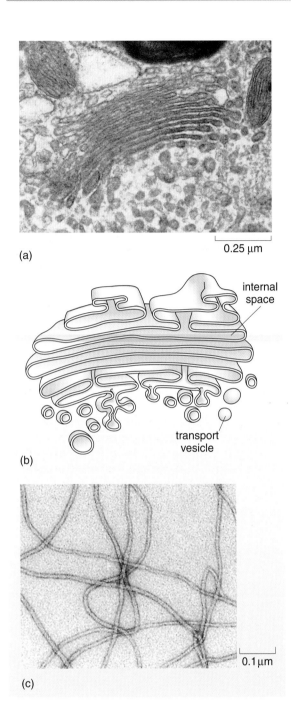

(a)

0.25 µm

internal space

transport vesicle

(b)

(c)

0.1 µm

Figure 3.4 (a) An electron micrograph of the Golgi apparatus, showing the stacks of membrane-bound sacs. (b) A three-dimensional representation of the Golgi apparatus. (c) An electron micrograph of intermediate filaments, consisting of tubes about 20 nm in diameter.

Protein-rich membranes are also a major constituent of a number of intracellular structures concerned with the manufacture and release of metabolic products. The most significant example is the Golgi apparatus, shown in Figure 3.4a and b, which consists of a stack of flattened, pancake-like sacks that form an important site for the packaging of proteins. Proteins manufactured elsewhere in the cell are often wrapped up in membranes for export from the Golgi apparatus, as the small vesicles that you see budding off the end of the sacs. Membranes are also key components of the scaffolding of **intermediate filaments**, which contribute to the internal architecture of individual cells (Figure 3.4c).

Again, the fact that particular proteins assemble within such membranes reflects their three-dimensional shape. For example, proteins of suitable shape create a pore within a cell membrane, to make the membrane selectively permeable. CFTR is just one such protein, described in some detail in Section 7.3 of Topic 6 in the context of cystic fibrosis. A variety of membrane proteins, in the form of pores and pumps, control the two-way flow of 'traffic' of molecules across the membrane – for example, promoting the influx of glucose, thereby supplying cells with a readily utilisable energy source. In general terms, the proteins within membranes are enormously important biological components. It is estimated that about 25% of all human genes code for membrane proteins.

One complex of membrane proteins is **ATP synthase**, currently the focus of much research attention – you might recall its brief mention by Sir Harry Kroto in the movie sequence that you studied in Chapter 1, who briefly explained its function. This is indeed a universally distributed membrane-bound structure, of central importance to the way organisms use energy.

■ Recall from earlier studies the function of ATP synthase.

▨ This enzyme promotes the formation of ATP, from ADP and Pi.

ATP has a central role in energy transfer, acting as a short-lived energy store that is constantly recycled. It is used to fuel the 'workings' of the body in such diverse forms as the secretion of manufactured products such as hormones from cells and the contraction of muscles. By one very approximate estimate, ATP is generated and utilised in our own bodies each day on a scale that is equivalent to our own body mass. Another expression of the key role of ATP is the speed with which death follows the poisoning (e.g. by cyanide) of the mitochondrial processes responsible for nearly all of its production.

Within the mitochondria, ATP synthase is the site for the movement of protons (i.e. H^+) across the mitochondrial membrane. As the protons pass through a pore that comprises the membrane-bound portion of the ATP synthase complex, a cyclic ring of proteins is now known to rotate in such a way that can result in the generation of ATP. Much the same type of membrane-bound enzyme occurs in the chloroplasts of plants; in some bacteria, a comparable mechanism, rotating at speeds in excess of 100 000 rpm, uses the movement of hydrogen ions to power the beating of a whip-like **flagellum**, which can propel the organism forward in a corkscrew-like locomotion. This structure is prominent in Figure 3.5, and consists of a lower, partly immobile component embedded in the membrane and an upper section that spans the membrane, which can rotate when powered by the movement of protons. The universality of rotary motors throughout living systems suggests that these complex nanomachines probably developed early in evolutionary history, as a highly efficient and adaptable means of managing energy in a biologically useful form, using the mechanical energy of rotation. Figure 3.5 also shows ATP synthase within the cell membrane of the bacterium *Escherichia coli*, considerably smaller than the flagellar motor alongside.

The techniques of nanotechnology are being increasingly used to study the action of a whole range of membrane proteins, including ATP synthase. It has proved possible to isolate and modify ATP synthase from cells and then link the enzyme complex with other molecular components by self-assembly. Such nanoscale devices can be fixed at one end to a flat substrate and at the other to a 'propeller' component, large enough to be observed microscopically. Once ATP is added as fuel, this ATP-synthase-based structure initiates rotation of the propeller.

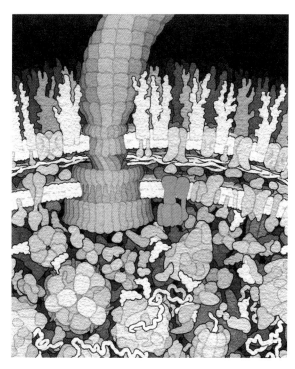

Figure 3.5   A computer-generated representation of an area adjacent to the outer cell membrane of the bacterium *E. coli*. Here, the cell membrane consists of two layers, an inner and outer. The large structure in pink is the flagellar motor of the bacterium, consisting of a number of complex proteins. The smaller rotary motor is ATP synthase, which spans the inner cell membrane and which can generate ATP.

So far, isolated molecular motors have largely been used to investigate how such devices function in the living cell. Questions of interest include how membrane-bound enzymes respond to particular stimuli, and how their functioning might be influenced by new pharmaceutical products. This is part of a more general trend of using the newly developed techniques of nanotechnology to increase our understanding of living systems, which is an aspect covered in the next section. Before moving on, it will be useful to remind you quickly of the ground covered so far.

Macromolecules, and also their manufacturing systems, such as the one concerned with protein synthesis, operate on the nanoscale. So do some entire organisms, such as viruses and some small bacteria. Proteins (and DNA) have the capacity to self-assemble, with particular sequences able to fold into stable, precise and predictable configurations. Lipids and proteins combine in the form of membrane proteins, often forming pores and pumps that control entry and exit of materials across the cell membrane. ATP-synthase is a widely distributed and ancient membrane-bound enzyme. In mitochondria, it displays an ability to convert chemical energy (in the form of movement of protons) into motion of a rotary motor, to produce ATP.

3.2 Using nanotechnology to understand and exploit biological systems

You know from Chapter 1 of so-called 'single-molecule techniques'. For example, the STM allows the direct observation of the behaviour of single molecules within biological systems. Indeed, a number of rapidly improving techniques, the details of which need not concern us, allow the monitoring of the motion of individual molecules, often tracking changes that occur over time. Recent advances in atomic force microscopy (another technique you read about in Chapter 1) have allowed researchers to follow conformational changes in proteins as they function. Individual proteins are first suitably 'anchored' and then stretched and unfolded; the process can be repeated after allowing the molecule to 'relax' back into its characteristically folded form. In such ways, the forces that hold proteins together are now beginning to be understood.

Somewhat similar techniques are being used to study the nanoscale 'motors' that contribute to the contraction of muscle fibres, each motor the size of a single protein molecule. This is an example of a linear motor, moving along a track in one dimension, in contrast to the rotary types typical of ATP synthase. To simplify greatly what happens at the molecular scale, muscle contraction depends on an interaction between one type of protein, consisting of **myosin**, and adjacent filaments of the protein **actin**. Figure 3.6 is a very simplistic representation of the steps involved. The diagram shows the critical steps of changes in the orientation of the head of the myosin molecule. A binding pocket exists on the myosin head, to which ATP first becomes bound, step (a). When ATP is broken down to form ADP and Pi, the myosin head is extended and sticks to the adjacent actin filament, steps (b) and (c). But once Pi leaves the myosin head, leaving just ADP in the binding pocket, the myosin undergoes a further substantial change in shape, step (d), which, rather like a lever, pulls the adjacent filament of actin in the direction of movement shown by the arrow, after which the myosin detaches. Huge numbers of myosin molecules acting in unison in such power strokes bring about muscular contraction; as you see, the role of ATP (and ADP and Pi) is critical to the process, helping create changes in the three-dimensional structure in myosin essential to the power stroke. This is a beautiful example of a highly efficient nanoscale motor in nature, and you can appreciate why researchers are so eager to use the techniques of nanotechnology to better understand the molecular mechanisms at work.

Another technique potentially of great significance to biology is the use of fluorescent nanoparticles to tag particular biological macromolecules in ways that will allow researchers to follow their fate within the cell. You know of quantum dots from Chapter 2 and their wide range of possible applications. Remember that when quantum dots of different size are excited by UV light, they will emit light of differing wavelength, i.e. colour. In a biological context, this property offers the prospect of labelling a very wide array of different proteins, for example, each with a quantum dot of a particular size. What is appealing is that, in principle, variations of this technique would allow quantum dots to be attached to a whole range of different macromolecules, in great numbers – some claim billions.

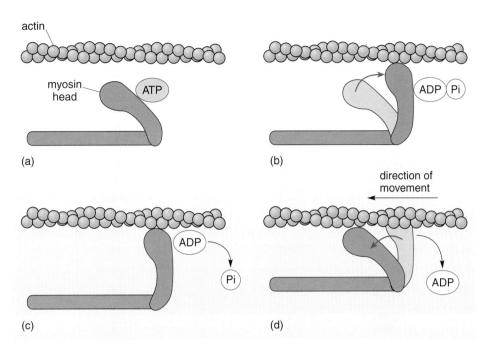

Figure 3.6 The myosin molecule, shown in purple, has a distinctive head, to which ATP is able to bind. (Two such heads are present, but only one is shown.) In this state (a), the myosin head does not bind to the adjacent actin molecules, but once ATP is broken down (to ADP and Pi) the head swings forward (about 5 nm) into the bent position (b), and the head binds strongly to the actin filament (c). Pi then leaves the myosin, which alters the orientation of the myosin head (d), creating the 'power stroke' that produces motion in the direction shown.

The use of quantum dots, therefore, offers great advantages over more conventional tracking techniques. Conventional tracking has often, for example, involved tacking fluorescent dyes to molecules. **Fluorescence** is the term that describes the emission of light from a substance (say a protein with a fluorescent component attached) when exposed to radiation of a shorter wavelength. But with the standard fluorescence techniques, each specific label has to be excited with a particular wavelength of light, so the potential for simultaneously tracking a great number of different fluorescent labels is very limited. Another problem is that such fluorescence tends to fade with time, and the numbers of distinct coloured components available for use is limited.

But with quantum dots, not only are their optical properties more stable, but multiple labelling becomes feasible, with simultaneous monitoring. You will appreciate from Table 3.1 that quantum dots, because they are roughly the same size as proteins, are good candidates to use for biological labelling. At present, techniques for specific labelling of this type are still in their infancy, but injecting quantum dots into whole cells is already yielding useful information. For example, when quantum dots are injected into the early embryo of an amphibian, they become located in particular cells that will go on to differentiate into specialised tissues such as nerve cells, liver and gut. The movements of such developing cells proved traceable by observing the quantum dots they contain. As the cells divide and differentiate during development, the quantum dots were readily identifiable, as they passed from the original cells to their many progeny.

It is clear, therefore, that techniques central to nanotechnology are increasing our knowledge of how biological systems work. The other side of the same coin is to exploit our knowledge of natural nanoscale systems to put together new nanoscale devices of potential practical value. For example, in the longer term, it may be possible to construct purpose-built artificial motors in the form of pores and pumps. Such devices might be able to pump ions or metabolites such as glucose against a concentration gradient. In other words, their use could promote the transport of a substance across a membrane from where it is in low concentration to where it is plentiful; the normal process of diffusion would result in net flow in the opposite direction, until an equilibrium was reached. Knowing what you do from Section 7.3 in Topic 6 of the importance, for example, of the balance of ions (such as Cl^-) either side of a cell membrane, you can see the potential for practical uses of membrane-based motors of this type.

The complexity of the ATP-synthase enzyme is such that any comparable artificial construct is indeed some way off. But simpler 'soft machines' of this type are being created, building on the capacity of biological nanoparticles for self-assembly. Different molecular components, some altered by the type of genetic manipulation you are familiar with from Topic 6, can be encouraged to stick together in an appropriate sequence.

No less remarkable, and potentially of great practical value, is the nanoscale machine developed largely by Bernard Yurke, working in Lucent Technologies' Bell Labs in the USA. Yurke reports that part of his inspiration for devising DNA motors was the realisation that molecular-scale protein motors in living organisms are responsible for muscular contraction – exactly the type you learnt about in Figure 3.6.

This machine, shown in Figure 3.7, is composed entirely of DNA and works by taking advantage of DNA base-pairing, forming what have been termed '**molecular tweezers**'. Two linked DNA strands (which come together by self-assembly, i.e. base pairing) form the 'arms' of this manufactured structure; see Figure 3.7a. For much of their length, each of the arms of the tweezers is double stranded, though a small central gap (arrowed) enables the arms to bend towards each other. However, each arm has a single-stranded extension, with exposed bases that are crucial to the process of closing.

■   Looking at Figure 3.7a, how might the addition of a third single-stranded DNA molecule bring the two arms together?

▨   If this third strand consists of bases that 'match' those of the exposed portions of the arms, then the structure would bend in the way shown in Figure 3.7b, where the third strand (called the 'fuel strand) is shown in (white).

A further clever piece of design relies on the presence in the closed position of still-unpaired bases on the fuel strand (see Figure 3.7c). So, were a fourth strand complementary to the entire length of the fuel strand added, this would pull the fuel strand off the tweezers, which would then open. Given that this added strand

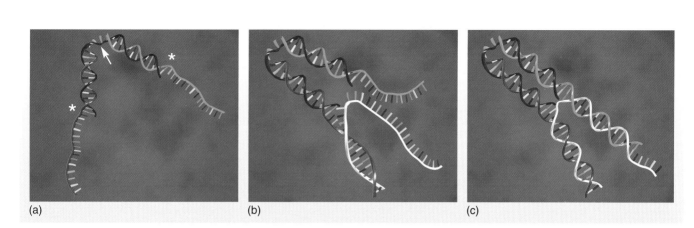

(a)   (b)   (c)

Figure 3.7   DNA nanomachine. Bases are represented by coloured bands. (a) Hinged double strand in open position, with unpaired sections exposed. Asterisks show the points of attachment of markers. (b) and (c) Addition of the fuel strand closes tweezers through base pairing with previously unpaired sections, bringing markers together. Though not shown on the diagram, the addition of another single strand, which pairs with the entire length of the fuel strand, will pull the latter off, reopening the tweezers.

has exposed bases that match those along the entire length of the fuel strand, this matching is the stronger, which explains why the fuel strand can be pulled off.

One problem that the researchers had to negotiate was how to detect opening and closure of the tweezers, given that the structure is too small to be visualised by existing microscopic techniques. The solution lay in attaching fluorescent molecules at the asterisked points shown in Figure 3.7a; when these marker molecules were closely adjacent (i.e. when the tweezers were closed) they fluoresced in a characteristic way under laser illumination. Calculations suggest that the strength of base pairing is such that the force exerted by nanoscale DNA tweezers of this type could be considerable by macromolecular standards – equivalent to that evident in the myosin motor mentioned already. Researchers see practical long-term applications for these remarkable nanomachines, e.g. in the construction of electronic circuits, involving what they foresee as the 'orderly addition of molecules'.

Activity 3.1

Allow 20 minutes

Recent research has resulted in the manufacture of a nanomachine that apparently 'walks' along a track composed of DNA. The research was summarised in the *New Scientist* in May 2004, and the diagram that accompanied that article is shown in Figure 3.8. For the moment, focus your attention on (a) and avoid looking through the remaining diagrams until prompted to, as you work through questions (i) to (v).

The track over which the machine 'walks' is composed of DNA, with spikes (or footholds) of unpaired DNA nucleotides that stick up; see Figure 3.8a. The legs of the machine (shown in purple) are double stranded for part of their lengths. They are linked at the top by a springy portion, linking left and right, and each has an unpaired 'sticky foot' in their lower portion.

(i) Why is the term 'sticky foot' appropriate for the lower part of each robot leg?

(ii) In Figure 3.8a, the right and left legs (in purple) are fixed to the walkway, at two points A and B. They are held there by two 'anchor' strands. How do the anchor strands achieve this, and is their make-up identical?

(iii) In Figure 3.8b, a free yellow strand attaches to the right anchor and strips it off walkway B. How can the yellow strand achieve this?

(iv) How can the foot that is now 'free' in (c) become fixed to foothold C, which would allow the nanorobot to step forward?

(v) How might the foot still attached to leg A be detached, to take a step forward onto foothold B?

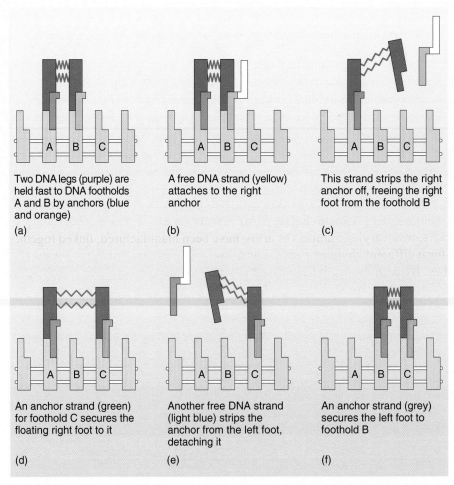

(a) Two DNA legs (purple) are held fast to DNA footholds A and B by anchors (blue and orange)

(b) A free DNA strand (yellow) attaches to the right anchor

(c) This strand strips the right anchor off, freeing the right foot from the foothold B

(d) An anchor strand (green) for foothold C secures the floating right foot to it

(e) Another free DNA strand (light blue) strips the anchor from the left foot, detaching it

(f) An anchor strand (grey) secures the left foot to foothold B

Figure 3.8   Diagram from *New Scientist* depicting the 'walking nanomachine'. The legs of the robot are shown in purple and the walkway, with labelled footholds, in pink. The 'anchor' and 'stripping' single-stranded DNA sequences are colour coded, as explained in the text and the activity answer.

Furthermore, base pairing within DNA allows quite complex machines to be fabricated and such shapes can be used as scaffold on which other materials can be deposited. It has been suggested that such techniques will eventually lead to the construction by self-assembly of an immense variety of intricate, precise, multidimensional nanoscale structures. There are a number of reasons that explain why DNA is regarded as such a promising nanoscale building material. First, as we have seen, the base-pairing rules can impart a high degree of control over the process – the specificity of pairing is of course exactly what makes DNA so well suited to its conventional role! Second, a huge range of base sequences can be constructed and manufactured 'to order'. There are added advantages; double helices of DNA are relatively stiff (at least in relatively short sequences), and they are stable under normal physiological conditions.

■ What natural enzymes exist that might be useful in helping to manipulate DNA to aid the construction of scaffolds?

■   You discovered in Topic 6 how restriction enzymes can 'snip' double-stranded DNA at characteristic sequences. There is also DNA ligase, which can help seal two single-stranded sequences, as long as their base sequences are complementary.

These enzymes provide the opportunity to construct **DNA scaffolds** on a modular basis. The individual building blocks, generally with four arms in the cross-shaped structure shown in Figure 3.9a, are brought together by specific and spontaneous base pairing. You will appreciate that if the base sequence of each particular 'arm' of the cross is unique, then the components can come together by self-assembly to form only one particular construction. Again, the clever element is that 'sticky ends' (Section 3.1 of Topic 6) are left on each of the branches; when the base sequences of these sticky ends are matched, the building blocks associate (Figure 3.9b). A wide variety of such DNA building blocks, with varying numbers of arms, have been manufactured, linked together to form different complex patterns and three-dimensional structures such as cubes. By such means, it has proved possible to create precise three-dimensional scaffolds using no more than simple one-dimensional rules of base pairing.

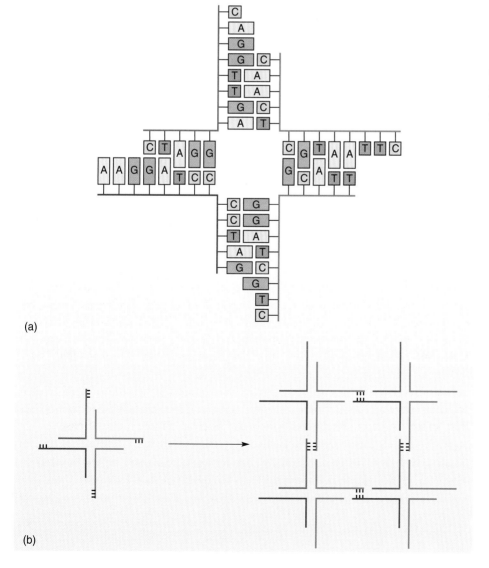

(a)

(b)

Figure 3.9   Large structures of DNA made by self-assembly. (a) A four-armed module is formed by specific base pairing. (b) Sticky ends ensure that an extended grid can be produced by self-assembly of identical four-armed modules.

Assembly of such scaffolds can be all the more efficiently done if the growing structure is 'fixed' on a bead and extra rigidity can be assured by adding extra linkages between strands. Short DNA loops can be added on; if restriction enzymes break off part of their structure, short 'hooks' are thereby exposed, which allow the addition of extra components.

DNA scaffolds are predicted to have great potential, but any practical application seems very far off. Some see DNA lattices as important devices in the future for storage of data and in the construction of molecular electronic devices. As with many such discoveries, practical applications remain at the level of informed and optimistic speculation.

The author David Goodsell describes our current state of development as 'toy problems', while stressing the much greater capabilities of these approaches for the future. The potential is all the greater because assembly of the type envisaged, notably that involving proteins, can be greatly enhanced by the use of genetic engineering techniques, in ways that reveal how these two fields of innovative research are moving ever closer together. Such techniques increase yet further the opportunities for adapting existing biological nanostructures, in ways that produce new constructs with novel functions.

Let us now briefly recap this section before moving on. You have learnt that the techniques of nanotechnology are revealing how single molecules function in living systems, especially changes in linear (e.g. myosin) and rotary (e.g. ATP-synthase) motors. Quantum dots are already proving an invaluable 'tracking' tool, in particular to trace the movement and function of proteins within cells. Furthermore, artificial DNA nanomachines (in the form of molecular tweezers) have been constructed, functioning via specific base pairing, which can perform mechanical work. DNA provides excellent material for scaffolding, where self-assembly through base pairing creates multidimensional constructs of a very precise and predetermined form. Finally, useful functions for artificial nanomachines have been predicted, but are as yet they are some way off, though the use of genetic manipulation is likely to hasten such outcomes.

3.3 Nanotechnology and medical science

You will be well aware by now of the contribution that nanotechnology may prove able to make to a very wide range of fields: information storage, lithography, environmental clean up, renewable energy. Some potential uses of nanotechnology have a less benevolent appeal. For example, it is predicted that a new generation of biological and chemical agents could be developed using nanotechnology that will be difficult to detect and deal with. Eric Drexler takes the view that nanotechnology will pose 'opportunities and dangers of first rank importance to the long-term security of the United States and the world'.

If, as many argue, nanotechnology is a 'mixed blessing', it is unsurprising that the potential benefits it offers medical science are very frequently brought to the fore. This section describes a number of promising new developments, many related to disease treatment and to drug delivery, that are in the earliest stages of development. As you read, bear in mind that making exaggerated claims of medical benefit is a tempting prospect for researchers, perhaps especially in a emerging area such as nanotechnology, anxious to establish its social value. In all fields of science, a very high fraction of medical 'hopes' for the future never become reality, falling at some stage in the continuum between the very expensive and time-consuming processes of laboratory development, animal testing and subsequent human trialling. Because of this high attrition, advances in medical science depend on huge numbers of promising possibilities competing for attention and funding.

Extravagant medical claims on behalf of nanotechnology are far from new and often merge into the realms of science fiction. Perhaps you can recall the 1966 film *Fantastic Voyage*. Its far-fetched storyline, involving the miniaturisation of a heroic crew of submariner scientists, entailed travelling through the bloodstream to remove a blood clot lodged within the brain and threatening the life of the patient. Nor have scientists been reluctant to speculate on wonder cures, from the earliest stages of the technology's development. Prior to the release of *Fantastic Voyage*, the Nobel laureate Richard Feynman said the following, in the now famous lecture mentioned in Chapter 1:

> A friend of mine (Albert R. Hibbs) suggests a very interesting possibility for relatively small machines. He says that, although it is a very wild idea, it would be interesting in surgery if you could swallow the surgeon. You put the mechanical surgeon inside the blood vessel and it goes into the heart and (looks) around. (Of course, the information has to be fed out.) It finds out which valve is the faulty one and takes a little knife and slices it out. Other small machines might be permanently incorporated in the body to assist some inadequately functioning organ.

Improbable though the more far-fetched of Feynman's scenarios remain, progress to date in this comparatively fledgling science is so rapid, and so much has been achieved, that we should use the word 'impossible' with great caution. And it is certainly true that some such medical applications are very much 'here and now', though these are far less spectacular. For example, nanocrystalline silver (unlike its bulk counterpart) has wide-ranging antimicrobial properties. Its application to wounds results in the release of ionic silver (over a sustained period of time) and it can act against more than 150 different bacterial pathogens. As well as 'incremental' improvements of this type, nanotechnology raises the prospect of revolutionary changes in health care. One such example has been touched on already in Chapter 1, which brings together the types of genetic engineering technique highlighted in Topic 6 and those discussed here. Section 1.2 mentioned that the detection of particular genetic sequences may prove possible using the small beams (cantilevers) that form the basis of the AFM. Some predict a scenario where tiny fluid samples can be diagnosed via cheap and reliable personal health monitors that, unlike today's instrumentation, are both affordable and portable. In such a way, information on genetic make-up (or on disease indictors) might become available 'in the home' without the need of specialised laboratory equipment or trained medical personnel.

It is perhaps the prospect of revolutionary change of this type that ensures this is a field where 'hype' is inevitably confused with 'hope'. An example is provided by the claims of the National Nanotechnology Initiative (NNI), which is the government-supported federal organisation that coordinates the funding and research activities for nanotechnology in the USA. In that role, the NNI is keen to promote the potential value of nanotechnology in the treatment of cancer. One source within the NNI has stated that 'it is conceivable that by 2015, our ability to detect and treat tumours in their first year of occurrence might totally eliminate suffering and death from cancer'.

Another example is from medical journalism, with the preface to an article on cancer research, which is pitched in ways that aim to grab the reader's attention:

> Here is the future of cancer as Naomi Halas sees it. During a cancer screening, your physician injects a gold-laden liquid into your bloodstream and shines an invisible light over your body for roughly 30 seconds. She turns to a computer monitor that displays a precision map of the size, shape and location of a newly budding tumour. The treatment? A hardier blast of the same invisible light. By the time you're back in your car, the growth is history.
>
> Kelleher (2003)

■ Are there elements in this scenario you find implausible, in that they are unlikely to be realisable within, say, the next 20 years or so?

▪ Talking in terms of *your* physician and by the time *you* are back in your car is a common journalistic device, to personalise the scenario, but to me there is an implication that such a scenario might occur within my own lifetime. There is also a rather incredible neatness about the scene that is described – with instant diagnosis and treatment – with no apparent confirmation, for example, that the young tumour has been eradicated. Finally, the assumption of the affordability of such routine treatment imparts a slightly unreal air to what is imagined.

As you will soon read, research involving nanoparticles of the type that Halas is concerned with is being very actively pursued and has many enthusiastic supporters. But, by most estimates, this is very 'long term'. However distant that time scale may seem, it is striking that the Royal Society and Royal Academy of Engineering 2004 report defines long term as 'over 20 years'. By that logic, wonder cures within one's own lifetime become more realisable, though the NNI's target date of 2015 remains overoptimistic. Indeed, responding to the NNI's claim, the Royal Society and Royal Academy of Engineering report concludes in appropriately sober tones: 'We have, however, seen no evidence to support the notion that nanotechnologies will eliminate cancer in the short-to-medium-term, and feel that such a claim demonstrated an over-simplistic view of the detection and treatment of cancer'. The report goes on to point out that although *some* measures based on nanotechnology *may* make some contribution to the detection and treatment of *some* forms of cancer, other steps (e.g. greater understanding of the environmental causes of cancer, better management of cancers) form no less important a role in reducing the impact of cancer in the foreseeable future.

All of the medical applications we will now look at are very much the province of the medium to long term. As you read them through, make a note of any major biological complications that you feel have to be negotiated and be alert to issues of safety throughout.

3.3.1 Peptide nanotubes

Some nanoscale constructions have a possible medical application in terms of their antibacterial potential, as revealed by the work of Reza Ghadiri of the Scripps Institute, California in the USA. The type of nanotube of interest here is distinct from the carbon nanotubes you read about in Chapter 2. These **peptide nanotubes** comprise rings of eight or so amino acids (so-called peptide rings) which are flat and, therefore, can be stacked one on top of another, with hydrogen bonds binding each ring to its neighbours. The side chains of the amino acids are oriented away from the rings, which means that a clear channel permeates through the core of the tube, as in Figure 3.10. Peptide rings of particular types are able to self-assemble in such a way that they associate very intimately with bacterial cell membranes, forming nanotubes tubes partly within those membranes, about 3 nm in diameter and 6 nm long. You will recall that transmembrane proteins (Figure 3.3) have the capacity to traverse the membrane, and those that form pores or pumps, for example, open up the membrane to two-way traffic, in a very regulated way. By contrast, peptide nanotubes have the capacity to disrupt and puncture these membranes, especially when they act in aggregate, with a resulting loss of a proportion of the cell's contents. This generally results in the death of bacteria within minutes, as opposed to the effect of more standard antibiotics, which normally take hours to exert their effects.

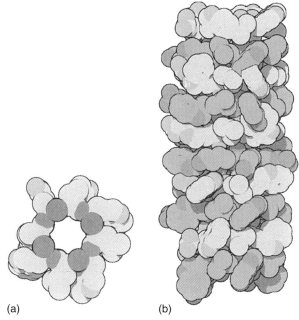

(a)   (b)

Figure 3.10 Cyclic peptides of the type shown individually in (a) can be assembled into nanotubes when stacked one upon the other (b). In (a), oxygen and nitrogen atoms form an inner ring; hydrogen bonds can link adjacent rings together. (b) resembles a 'rolled-up' version of the β-sheet shown in Figure 3.2b. All the side groups project outward, ensuring a smooth central hole.

■   What problems, of the type associated with more conventional antibiotics might be associated with the use of peptide nanotubes?

▨   The development of resistance is a major problem, as the chemical nature of the bacterial coat changes via mutation. For instance, **methicillin-resistant *Staphylococcus aureus* (MRSA)**, or more commonly, 'hospital superbug', is an increasingly threatening presence in UK hospitals, with 15% of infections resulting in death. Another problem to be negotiated is how to ensure the specificity of action of peptide nanotubes, in this instance to act against the membranes of particular harmful bacteria and not to act, for example, on those of red blood cells.

One fruitful approach to the problem of specificity has been to try out variations by altering the side chains of the amino acids that comprise peptide nanotubes. Such design differences do not have a profound influence on the shape and dimensions of the tube, but they influence the interactivity of the nanotubes with biological membranes. Several initial trials have involved the construction of nanotubes against MRSA. *In vitro*, they proved effective against MRSA, but not against *E. coli*. Similar success was evident *in vivo*: when mice were injected with lethal doses of MRSA, together with specifically designed peptide nanotubes, they resisted the infection. The development of bacterial strains that are resistant to such drugs is a possibility, but the hope is that as the character of bacterial membranes changes, as is inevitable under the combined influence of mutation and natural selection, nanotubes can be swiftly changed to keep pace. Peptide nanotubes can be synthesised relatively easily (and cheaply), so extensive 'libraries' of such nanotubes could perhaps be developed, each geared to particular bacterial strains.

So, do cyclic peptides offer advantages over more standard antibiotics? Most conventional antibiotics depend on their action on specific interactions at particular molecular sites, on the basis of precise molecular recognition. This means that alteration of a few amino acids within a particular receptor will often mean that binding between antibiotic and receptor cannot occur. Changes on the same small scale might render a particular enzyme unsusceptible to an antibiotic. So, minor alterations in molecular structure (routine events in bacteria) can lead to the development of bacterial resistance, with that trait rapidly becoming more prevalent in the bacterial populations as a whole, via natural selection. The evidence to date suggests that the mode of action of peptide nanotubes is maintained despite significant biochemical changes within bacteria, suggesting that resistance might be less of a problem. Of course, bacteria could develop resistance to peptide nanotubes if they were to change the composition of their membranes very significantly. But this is likely to occur only via the modification of a number of biosynthetic pathways, as opposed to relatively minor mutations that are all that is required for conventional antibiotic resistance. Complex changes on this scale would seem more improbable than the simple mutation of an enzyme or the modification of the chemical make-up of a particular receptor.

3.3.2 An artificial pancreas

You will know about diabetes and the role of insulin from Topic 6. Many of the different types of artificial pancreas currently being developed involve implanting insulin-producing pancreatic cells within the body of diabetic patients in ways that allow the release of insulin.

■ Why is it impractical simply to inject such cells into the recipient's pancreas?

▨ Foreign cells of this type would be destroyed by the antibodies that constitute the body's defence mechanism to combat invasion. (You will recall from Topic 6 that this would not happen with cells cloned from the patient's own cells.)

One option is to suppress a patient's immune system, but the powerful drugs used to this effect generally have side-effects and leave the patient very vulnerable to what would normally be minor infections. One 'nanotechnological' approach is being undertaken at the time of writing (2006) by Tejal Desai, partly within Boston University and partly within a private bio-engineering company. Her work involves the development of small silicone capsules, each about 2 mm in diameter, that are filled with cells obtained from the mouse pancreas; the walls of the capsules are permeated with nanoscale pores, approximately 7 nm across.

The pores are created by the process of photolithography, which, as you know from Section 2.1.2, is a standard device for microchip manufacturing. Pores of this magnitude are big enough to allow the passage of insulin, which is a moderately sized protein (Table 3.1), and most importantly of glucose, which is required to trigger the release of insulin from the pancreatic cells. But they will not permit the passage of more sizeable antibodies, notably **immunoglobulin G (IgG)**, which ought to protect the transplanted cells from attack and destruction.

At the time of writing (early 2006), experiments have involved *in vitro* work, plus *in vivo* approaches involving the implantation of the capsules into rats that were diabetic. Results to date are described as 'promising'.

■ What questions do you think need to be addressed before this technique could become available for routine use in people with diabetes?

▨ Some problems relate to ensuring effectiveness. For example, how many capsules would have to be injected? Would more capsules have to be added periodically? Might pores become 'clogged up' if cells such as macrophages started to invade the capsules? Other questions relate to safety; would particles of this magnitude have any untoward effect on the body? (This is a point to be picked up later on.)

Question 3.2

A question often asked of new technologies is whether they are *needed*. Could they prove to be better than what is available at present? For each of the medical techniques outlined so far, i.e. peptide nanotubes and Desai's artificial pancreas, summarise the potential advantages over and above more conventional approaches.

3.3.3 Nanoparticles and drug delivery

A number of drug delivery mechanisms on the nanoscale are under investigation and we will look at some representative examples in this section.

Quantum dots

Some approaches use the properties of quantum dots, mentioned in Section 3.2 and first introduced in Section 2.1.4. For example, crystals of the semiconductor cadmium selenide are encapsulated within a coating in which is embedded the cancer drug Taxol (Figure 3.11a). Within the same coat are antibodies that are specific to folic acid receptors present on the surface of cancer cells in numbers about a 1000-fold greater than in normal cells. When these nanoparticles, just a few nanometres in diameter, are injected into mice, they locate preferentially on cancer cells, via the association of folic acid receptors and antibodies (Figure 3.11b), and are then taken into the malfunctioning cells (Figure 3.11c). Then the animal is illuminated with infrared light. Such wavelengths penetrate living tissue comparatively easily, so if the tumour is relatively superficial (e.g. cancer of breast tissue), then sufficient is absorbed to excite the nanoparticle.

What is crucial here is that sufficient energy is released by the process to cleave the bonds that bind Taxol to the coat of the nanoparticle, delivering the drug to the required site of action (Figure 3.11d), resulting in the death of at least a proportion of the cancer cells. Much the same effect was apparent when tested against human prostate cancers that had been introduced into mice, implying that in years to come the procedure could prove of medical value.

Figure 3.11 Schematic representation of how quantum dots could achieve targeted drug delivery. (a) Coated quantum dot, with embedded Taxol and specific antibodies. (b)–(d) Stages in the location of cancer cells, excitation by infrared light and intracellular Taxol release.

(a) Nanoparticles are injected. Antibody targets them to cancer cells.

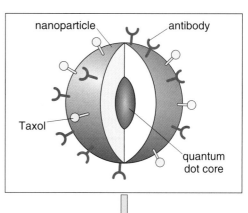

(b) Antibody attaches to cancer cell.

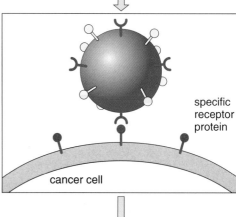

(c) Cancer cell takes in the nanoparticle.

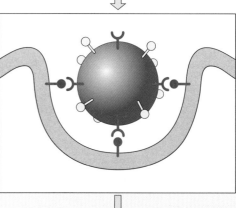

(d) Infrared light is shone on the suspected site of tumour. Quantum dot releases photons that liberate Taxol molecules bound to nanoparticle. Taxol molecules destroy tumour.

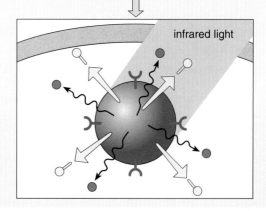

Nanoshells

Nanoshells were the focus of the speculative news article on the work of Naomi Halas highlighted earlier in Section 3.3. As with quantum dots, in the long term they offer the prospect of flexible and responsive drug delivery. Nanoshells take the form of a small core component, between 50 and 1000 nm in diameter. Most are composed of silica, enveloped by a membrane made of metal – normally gold, because it is biologically unreactive. Figure 3.12 is a computer simulation of some of the construction steps involved.

Figure 3.12   Simulation of nanoshell construction, with an overlayer of gold enveloping an inner silica core.

Nanoshells can be injected into tumour cells, either human tissue growing in culture or into tumours of living mice. They absorb light of a precise wavelength, generally in the infrared range, which is converted to heat, which spreads to the adjacent cancer cells. Even mild heating (to about 55 °C) disrupts the normal transport across the outer membrane of cancer cells, often leading to cell death. The gold surfaces of the nanoshells are also sites where antibodies can be attached, again binding to the folic acid receptors of cancer cells. Furthermore, nanoshells can also be joined to drug-containing capsules, where the outer layer is a heat-sensitive polymer. When the attached nanoshell is activated, the capsule is prompted to release its contents.

■   What advantages might drug delivery by nanoshells (and quantum dots) have over tried-and-tested routes, such as administration orally or by injection?

▨   Many drugs cannot survive passage through the stomach; some drugs (such as anticancer drugs) have unpleasant side effects, in large part because they influence tissues other than the target site. Nanoshells offer the prospect of targeting particular sites in a much more specific way, perhaps reducing the amount of drug needed. (There are also potential advantages in that nanoshells can deliver highly insoluble drugs that couldn't be conveyed via the bloodstream.)

In vivo experiments, creating local heating of mice tumours via nanoshells, were reported by Naomi Halas as '100% successful', adding 'that was really stunning for us, because it worked so extraordinarily well the first time it was done – and that, in experimental science, never happens … It's caused a tremendous amount of excitement within our own work and within the company'. Something of Halas's strong motivation emerges from her comments:

> I get contacted quite a bit about nanoshell cancer therapy. In fact, just in the last 24 hours I've had three contacts about this, ranging from someone's spouse who is in critical condition to someone's nine-year-old kitty who has a lot of tumours. It's an amazing experience to have these communications. I try whenever I can to reply to people who do make contact with me, even though there are no human trials currently – they are in the planning stages – and I'm not a part of the human trials in terms of any co-ordination – this is all being done through the company.
>
> Yes, I really look forward tremendously to the day when we can point to people – I want to know their names, their faces – who have been cured with nanoshell-assisted cancer therapy. That's a day that I cannot wait for. I'm very eager to see the smiling faces of the survivors and family members when that happens.

Interview with Naomi Halas, *Nova Science Now*, April 2005

Activity 3.2

Allow 15 minutes

Read the following section from an article entitled 'Metal nanoshells; cure or curse?' written coincidentally by a very vocal opponent of GM technology – Mae-Wan Ho, formerly Reader in Biology at The Open University. After an early section that described the potential of nanoshells in fairly neutral tones, including Halas's work, the article then strikes a more critical note:

> But are the nanoshells safe? They are non-biodegradable and have enhanced catalytic capabilities. What happens to the nanoshells in the dead cells when they are cleared by the immune system? What effects do they have on the health of the patient in the long term? What are the wider environmental impacts when these nanoshells are discharged or released? None of these questions have been addressed.
>
> It is clear that enthusiasm to exploit the remarkable properties of metal nanoshells and other nanoparticles have run far ahead of any safety concerns. It is time for responsible scientists to impose a moratorium on research and development until proper safeguards are put in place.
>
> (Ho, 2004)

Would you support this plea for a moratorium on research? Which case do you find the stronger: Halas's arguments to see her work come to fruition, or Mae-Wan Ho's concern about safety? As you think about safety issues, you may find it useful to look back at Section 1.3.5, which briefly considered the health risks of inhaled nanoparticles.

In more sober language, a recent report from the UK's Health and Safety Executive expresses much the same concern about safety issues:

> The duration of the deposits there [i.e. the liver], the potential harm they may trigger, and what dose might cause a harmful effect have not yet been exhaustively examined. Other diseases of the liver indicate, however, that even the accumulation of completely harmless substances can impair and damage the functioning of that organ.
>
> (Health and Safety Executive, 2004)

Dendrimers

Another class of nanoparticles with a great potential for drug delivery are organic dendrimers, which you read about in Section 1.4.3. You will recall that these are highly branched polymers, about the size of proteins, anywhere between 2 and 20 nm in size. But their structure is less easily altered than the proteins we have mentioned, given that the branches are held together by strong bonds; for practical purposes, this gives dendrimers a welcome robustness and permanence.

Dendrimers acquire their globular shape in a way quite unlike proteins, the most familiar of which, as you know, arise by folding of a linear chainlike structure. Figure 1.23 shows just a few examples of dendrimer design.

■ Recall from Section 1.4.3 how dendrimers grow.

▨ They are built up progressively, in cycles, with reactive terminal groups of one 'generation' acting as a foundation for the next.

During their synthesis, the reactive terminal groups can be fashioned such that they link to particular chemical groups. These tips can be linked to antibodies, or to particular metal atoms. Just as conveniently, you will recall that the cavities within dendrimers can readily accommodate 'guest' molecules – therapeutic agents perhaps. Indeed, dendrimers appear to have the capacity to cross cell membranes relatively easily, which is one reason why they are seen by some researchers as a better option for the 'smuggling' of DNA fragments into cells than some of the vectors you read about in Topic 6, notably the engineered viruses. A lot of work is currently underway on how dendrimers might help combat cancer. A single dendrimer type might carry a battery of components that could help locate, label, visualise (by fluorescence) and treat (i.e. by drug delivery) cancer cells. There is hope that a fifth agent could be added on, as a signal 'mission accomplished' molecule, released as cancerous cells are destroyed.

■ Do you think constructing a complex dendrimer of this type, step by step, is a very efficient way of putting together an agent such as this, or would it be more useful to construct a range of single-function dendrimers and apply them separately?

▨ An all-purpose dendrimer of this type would be advantageous because so much is delivered in one step. The problem is that creating such a dendrimer is an extraordinarily complex and time-consuming procedure.

One problem inherent in the synthesis of dendrimers is that there is some loss of yield in each of a succession of steps, so the amount of usable product at the end is a small fraction of the starting component. Thus, a rather more practical approach is the construction of two different dendrimers, one perhaps with 'add-ons' helping to visualise cancerous cells via fluorescence, the other containing folic acid. Each type of dendrimer carries a characteristic single-stranded DNA sequence on its outer surface, about 30 or so bases long. These DNA sequences are complementary, so when the two dendrimers types are mixed in solution, a process of self-assembly will ensue, such that the complex formed has the folic acid-related antibody component at one end and the fluorescence tag at the other. These complexes indeed proved able to locate cancer cells specifically, as indicated by their fluorescence. Figure 3.13 is a computer-generated impression of several pairs of conjoined dendrimers, showing the ability of the antibody component to locate on the cell surface. More complex packages are being designed, with further complementary DNA sections linked to dendrimers containing anticancer drugs.

Dendrimers have also been tagged with iron oxide (producing what have been called 'magnetodendrimers') and then grown in culture with stem cells that were derived from rat brain tissue (i.e. neuronal

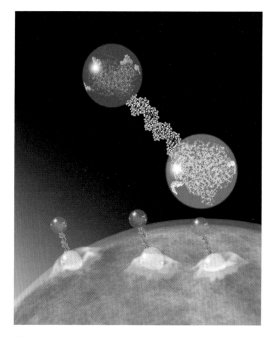

Figure 3.13 Multifunctional dendrimers linked by a DNA section formed by the base pairing between two complementary strands.

stem cells). The iron oxide-tagged stem cells that resulted were then injected into the brains of rats. It was relatively straightforward then to follow the movement and fate of the injected stem cells in the living animal, using an imaging technique (magnetic resonance) to detect the iron oxide in the dendrimers. Given the current interest in using stem cells for medical treatments (Section 6.1, Topic 6) you can appreciate the value of such information on stem cell behaviour. Magnetic nanoparticles are also being used to construct diagnostic probes; when these are linked chemically to particular antibodies, they can be used, for example, to detect the beginnings of cellular malfunction after their injection into a test animal.

A medical role for buckminsterfullerene?

You will recall mention of buckminsterfullerene in Chapter 1 (Box 1.1), often referred to as a C_{60} buckyball. This molecule has attained iconic status in the world of nanotechnology, given the impact of its discovery in 1985. However, up until recently, although techniques have been developed for its production in sizeable quantities, this enormously important molecule had great symbolic significance for nanotechnology but had limited utility. Now there is a great deal of interest in its scaffolding role, as a vehicle to convey other molecules to particular sites of action. The chemist Stephen Wilson, based at New York University, was co-founder of the fledgling nanotechnology company C-Sixty, and one of their most significant products to date has been an anti-AIDS drug that comprises C_{60} with dendrimers. What is especially significant is that this complex is able to bind to the viral enzyme reverse transcriptase (Box 2.1 of Topic 6). This enzyme 'reverse' transcribes the RNA of the virus into DNA, which then becomes the template used by the *host* cell for the replication of the virus. The fullerene-based drug is able to prevent replication because of its ability to bind to the active site of the reverse transcriptase enzyme, thereby blocking its activity. For reasons that are not yet clear, it also seems to kill both the HIV viruses (HIV-1 and HIV-2), despite the differences between them. The class of anti-AIDS drugs called protease inhibitors, work in a comparable way on reverse transcriptase, binding chemically to the active site of the enzyme. By contrast, the fullerene appears to form a mechanical plug adjacent to the active site and, therefore, is less sensitive to minor alterations in the chemical make-up at the active site.

- ■ Protease-resistant strains of AIDS are beginning to emerge, and the explanation seems to be minor changes in the amino acid sequence close to the active site of reverse transcriptase. Why might fullerene-based blockage be less susceptible to modification of this type?

- ▨ The argument here is similar to that you encountered in Section 3.3.1 on peptide nanotubes. Small changes in amino acid sequence might be enough to alter the shape and binding potential of enzymes, but large-scale changes may be needed to prevent a physical blockage, and these are less likely to come about.

3.3.4 Nanoparticles and the blood–brain barrier

The brain is well protected from invasion by potentially harmful chemicals within the blood by special properties of the cells that line the extensive and very thin capillary vessels that permeate brain tissue. In particular, the junctions between these lining cells are especially 'tight', with just a few substances (caffeine and alcohol being all-too-familiar examples) able to get through. Finding ways of circumventing the **blood–brain barrier** is a key target for the pharmaceutical industry, with the object of targeting brain tissue with drugs that might help alleviate or reverse malfunctions of nervous tissue, such as Alzheimer's or tumour growth. There is much interest in using nanoparticles as 'carriers' to traverse the blood–brain barrier, with drugs attached. Why nanoparticles have this unusual ability to circumvent the 'tightness' of the blood–brain barrier is not yet fully understood. But what is sobering is that, so far, promising results obtained *in vitro* are not readily reproduced *in vivo*. One problem is that the nanoparticles–drug complexes have to be injected intravenously. This means that, *en route* to the brain, they pass through the liver, which can remove and/or inactivate them. Thus, the great hopes for the technique, sometimes expressed euphorically in terms of 'enabling treatment of illnesses hitherto considered incurable – bringing hope to terminally ill patients', may be a very long way from realisation.

Activity 3.3

Allow 30 minutes

Up to now, section summaries have been supplied. Writing your own summary is often an excellent means of identifying and remembering key points. Write a summary of Section 3.3 and compare what you write with the comments provided. To help the comparison, produce a numbered list of about eight key points.

3.4 How and in whose interests is nanotechnology going to develop?

The examples of nanotechnology described in this chapter raise a number of broader issues, many of them ethical in nature. There has been the opportunity to discuss only a small number of the many potential applications of nanotechnology to medical science: important areas such as its role in tissue engineering (replacing worn or damaged bone, for example) have not been mentioned. But the medical examples have provided sufficient food for thought with respect to issues such as who 'controls' nanotechnology and whose interests are served by current advances.

It is self-evident from what has been said that nanotechnology has huge potential. Some claim the field is moving ahead via (in the words of one critic) 'the unquestioning exploitation of new science and technology for wealth creation and exploitation'. Indeed, the enthusiastic supporters of nanotechnology do see this field as the harbinger of the 'next industrial revolution'. Just as was the case with the GM controversy, a polarised debate is discernable. One camp is urging rapid growth of nanotechnology – bringing claimed benefits of economic growth and new forms of medical treatment – and those on the other side are keen to see that advances are in some way controlled and perhaps halted altogether.

Nanotechnology is certainly nowadays 'big business' and academic communities are strongly being urged to maximise opportunities for commercial exploitation. In the UK for example, the Engineering and Physical Research Council, one of the most significant funders of research in the physical sciences, has adopted a strategic aim of a research culture by 2007 that includes '50% industrial participation'. Again, in the UK and elsewhere, there are large numbers of companies in the field of medical applications alone, and Table 3.2 includes details of just a few such companies, as of 2004. You will see that many of the techniques mentioned so far are under development, notably in relation to drug delivery. The list could be a great deal longer; the examples shown simply aim to illustrate the diversity of approaches that have commercial application.

Table 3.2 Examples of companies commercialising nanoparticles for biological and medical applications, based on information from a 2004 publication

Company	Major area of activity
Advectus Life Sciences Inc.	drug delivery, especially anti-tumour drugs across the blood–brain barrier
Alnis Biosciences, Inc.	biodegradable nanoparticles for drug delivery
BASF	toothpaste; nanoparticles to improve dental surface
Biophan Technologies, Inc.	carbon composite materials for shielding in magnetic resonance imaging equipment
Evident Technologies	luminescent biomarkers, in particular quantum dots
Immunicon	tracking and separating different cell types, using nanoparticles with magnetic core and coated antibodies
KES Science and Technology, Inc.	nano-TiO_2 to destroy airborne pathogens
NanoCarrier Co. Ltd	drug delivery, especially micellar nanoparticles for drug, protein, DNA delivery
NanoPharm, AG	nanoparticles to cross blood–brain barrier
Nanoprobes	gold nanoparticles for biological markers
NanoMed Pharmaceutical, Inc.	drug delivery
Oxonica Ltd	sunscreens; nanoparticles to absorb UV
PSiVIDA Ltd	tissue engineering; material properties of nanostructured porous silicone
Smith & Nephew	bandages, using nanocrystal silver
Quantum Dot Corporation	luminescent biomarkers

Activity 3.4

Allow 20 minutes

Look back at the example of medical applications mentioned in this chapter. Is there evidence of the ways in which these key researchers are moving their developments forwards? In particular, to what extent are private companies involved?

This is an area very different from what might be termed more traditional academic research. Take the example of a leading US researcher, Paul Alivisatos, writing in *Scientific American* on the prospects of nanotechnology in the field of medicine (Alivisatos, 2001). Part of his article addresses the idea of using quantum dots, utilising some of the approaches that we have touched on. Alivisatos mentions the company Quantum Dot Corporation (see Table 3.2) 'leading the push towards commercialising this technology', in collaboration with colleagues at other US/foreign universities. He adds 'I helped to found this company, so my assessment may be biased, but I view the prospects for quantum dots as, well, bright'. This author has the wisdom to declare his involvement, but such associations may not always be so forthcoming. When I last visited the website linked with this company, I found an interesting mix of academic information about quantum dots and commercial incitement, presented in a way that goes against more traditional views of academic impartiality.

It is striking that many such companies have small 'start-up' status, spun-off from the initial work of academic researchers, with the likelihood that many will be taken over by the big pharmaceutical companies. One consequence of this heavy commercialisation of nanotechnology is that not all the work underway is in the public domain, with patents restricting what can be revealed. A strong argument made by critics of existing arrangements is that, again rather like GM, such patents have the potential to act against the public good, stifling innovation and reinforcing existing inequalities.

A key concern that will weigh in the public's mind is 'who benefits from nanotechnology?' Some claim that the early examples of nanoproducts reflect a 'rich-world agenda' – better suncreams, computers and tennis racquets for example. Some such critics claim that a 'nanodivide' between 'winners' and 'losers' is already emerging, with much the same language that is evoked in other contexts, including that of the climate-change debate that you are familiar with from Topic 5. The sharpest critics of nanotechnology, therefore, argue that progress on nanotechnology is acting to the detriment of wider society. And yet Douglas Parr of Greenpeace highlights the potentially beneficial applications of nanotechnologies for the developing world and for the environment, for example by reducing CO_2 emissions through improving renewable energy technology, expressing concern that nanotechnology could become another 'opportunity lost' for developing countries. In the field of medicine, it is likely, for example, that the use of nanotechnology will herald a new era of reliable diagnostic kits, more portable and more cost-effective than existing systems.

Other commentators are much more pessimistic in their assessment, drawing attention to 'unanticipated consequences' and to 'revenge effects'. Such critics point out that unforeseen consequences have affected society in the past, creating change in ways not always intended – and make the powerful point that we can expect no less from nanotechnology. And yet, as this chapter has shown, the realisation of some of the hoped-for benefits from the technique stand a good chance of producing a most welcome bounty for human health. But quite how nanotechnology might move forward in a way mindful of the 'public good' is by no means clear, which is one of the areas that Chapter 4 touches on.

Summary of Chapter 3

Living cells contain natural macromolecules, such as DNA, and manufacturing systems, such as ribosomes, constructed on the nanoscale. Self-assembly is a key property of many such naturally occurring nanoparticles, notably proteins, which generally adopt a characteristic three dimensional shape that reflects the nature and linear sequence of the amino acid constituents. Membrane-bound proteins often take the form of pores and pumps, controlling the two-way transfer of material across the lipid-based cell membrane. The enzyme ATP synthase is located within the mitochondrial membrane, and part of its structure includes a rotary motor, the rotation of which is linked with the generation of ATP.

The mechanisms that power linear and rotary motors are being studied, using the techniques of nanotechnology. The role of proteins and other cellular components is being investigated using nanotechnological tools, notably quantum dots. Artificial DNA nanomachines can now be constructed, such as DNA tweezers and a walking device, both of which depend on specific base pairing for their action. Scaffolds of DNA are being manufactured, to precise and consistent forms.

Nanotechnology promises much in the field of medicine, though relatively few of the products under development are likely to prove safe, effective and an improvement on more traditional methods. Cyclic peptides have promise as antibacterial agents, perhaps less susceptible to the development of resistance. Nanoshells and dendrimers are being actively researched as devices for delivering drugs, ideally combined with diagnostic and/or visualisation functions. Great though the hopes are for such techniques, their effects on cellular function more generally have not as yet been investigated – little is known about possible toxicity and how/whether such nanoparticles would be cleared from the body. The proven ability of nanoparticles to cross the blood–brain barrier offers a potential means of accessing and treating debilitating neural diseases.

The commercial potential of nanotechnology accounts in part for its enthusiastic support, from academia and from governments alike. The extent to which developments in future can be geared most effectively to the public good is very unclear, though there is widespread concern to avoid the public disquiet that accompanied earlier public controversies, such as GM food (Topic 6).

Questions for Chapter 3

Question 3.3

Can self-assembly occur between different types of biological macromolecule?

Question 3.4

Consider the quote, 'biology never invented … the wheel'. On what grounds could this assertion be questioned?

Question 3.5

By what means are the 'sticky ends' likely to be produced in the creation of DNA scaffolds and in what way might they be useful?

Question 3.6

Peptide nanotubes appear to exert their effects over a large proportion of the bacterial cell membrane. How is this different from more conventional antibiotics and why is this significant?

Question 3.7

How is specific base pairing used to put together nanoparticles able to deliver drugs to specific target sites such as cancer cells?

Question 3.8

Is it likely that the health risks posed by the use of nanoparticles, such as dendrimers, for drug delivery would include concerns about them becoming stuck within the fine capillary blood vessels that permeate all tissue?

Question 3.9

(a) One particular nanoparticle has been described by a researcher as a 'tunable, precision-guided thermal bomb'. Identify the type of nanoparticle described and assess whether the description is accurate.

(b) Dendrimers (and nanoshells) have been described as non-toxic. Is this a fair comment and would such a property be an advantage over the more conventional form of treatment for cancer, notably chemotherapy?

Question 3.10

Describe what is meant by the blood–brain barrier. What are the implications of nanoparticles being able to traverse it?

Chapter 4

Nanotechnology futures

Many of the debates surrounding issues in nanotechnology and its future direction are rather different in nature from those you have considered in the other topics of this course. In part, this arises from the enormous spectrum of science (and scientists) involved in the field: scientists with different training, methodologies, perspectives and goals can bring very different visions of what a nanotechnology revolution might entail. This can make it difficult for some of the scientists involved to communicate effectively with one another, let alone with industry, government and the public. Also, in comparison with the other topics covered in this course, nanotechnology is arguably still at an almost embryonic stage. In the uncertainty about how it will develop lies much of its excitement, but also many of the concerns associated with it. Although most of what nanotechnology has delivered so far in terms of useful applications (at least at the time of writing in 2006) represents incremental progression from earlier technologies, rather than totally new points of departure, there has been huge economic investment in the promise of a scientific and technological revolution. This has led to an understandable anxiety to jump on the bandwagon, which may explain some of the exaggerated claims made for nanotechnology products, not only in the media, but also by commercial companies and sometimes by scientists themselves. The flip side of that coin is a set of more or less extreme disaster scenarios advanced by some individuals and pressure groups, concerned that the lure of economic benefit may mean that potential risks are not adequately considered. However, lessons were learned (perhaps especially in the UK) from the experience of adverse public reaction to some aspects of GM, and it was seen as desirable that nanoscience, commercial interests and public opinion should not remain entirely detached from one another. This led both national governments and non-governmental organisations (NGOs) to want to initiate discussions about nanotechnology with different interest groups while it was still very much an emerging field, so as to inform decision making. Professional bodies in several countries held debates about the ethical issues surrounding nanotechnology. Questions of regulation and legislation were raised. Ideas for engaging the public in the debates have included new types of participation process, such as NanoJury UK, which was held over the summer of 2005.

This chapter briefly examines a few of these different debates about the future of nanotechnology. As you read, it is worth bearing in mind that during the life of the course some of these debates may move forward significantly and new issues are likely to emerge.

4.1 The debate about assemblers

If Feynman can fairly be said to have been the first scientist to envision a nanotechnology revolution, as already described in earlier chapters, the most prominent advocate of radical new approaches to nanoscale technology was K. Eric Drexler, who in 1986 wrote a popular science book on the subject, entitled *Engines of Creation: the coming era of nanotechnology*. In this volume, he first set out his vision for 'molecular manufacturing', which he described by analogy with conventional computer-controlled manufacturing. Working from the perspective of a systems engineer, he envisaged structures being assembled 'with atomic precision', using mechanical controls whose parts (robotic arms, for example) would be of nanometre scale. These machines could in theory be programmed to build any chemically stable structure or device from small molecular building blocks and were therefore called **assemblers**. Drexler has continued to develop these ideas for 'productive nanosystems', with technical and theoretical analyses. He also founded the Foresight Institute, a US non-profit organisation, which promotes his radical vision.

A major element in Drexler's early thinking was that such robotic assemblers – often called nanorobots or nanobots – would have one feature that has no analogue in macroscopic machines: they would also have the capacity to produce copies of themselves. In other words, assemblers would be able to self-replicate, a capacity that is familiar in the context of whole cells and of DNA (and also of computer viruses, which replicate strings of bits), but not normally associated with machines. However, Drexler recognised the potential threat of such replication. In *Engines of Creation* he described how assembler-based replicators (the US spelling has been reproduced here)

> could spread like blowing pollen, replicate swiftly and reduce the biosphere to dust in a matter of days … Among the cognoscenti of nanotechnology, this threat has become known as the 'gray goo problem'. Though masses of uncontrolled replicators need not be gray or gooey, the term 'gray goo' emphasizes that replicators able to obliterate life might be less inspiring than a single species of crabgrass … The gray goo threat makes one thing perfectly clear: we cannot afford certain kinds of accidents with replicating assemblers.
>
> (Drexler, 1986)

This theme is a fruitful one for writers of science fiction, but has also engaged the attention of others concerned about the risk and ethical issues associated with proliferating nanobots. In 2000, Bill Joy, co-founder of Sun Microsystems and at that time its chief scientist, wrote a famous article in the magazine *Wired*, in which he discussed the dangers to the human race of technologies involving robotics, genetic engineering and nanotechnology. Joy spoke from personal conviction, but also as someone who understood technology and the nature of complex systems. He could envisage scenarios, including the possibility of dangerous replicators escaping from laboratories or even being created with malicious intent, in which 'the future doesn't need us'; human beings could become an endangered species. The *Wired* article was important in bringing the issue of self-replicating assemblers into public view, not least because it explicitly linked the risks of genetic engineering with those of nanotechnology. There had already been a public backlash against GM in the UK and elsewhere in Europe. Was there going to be a similar reaction to nanotechnology before it was even fully fledged?

From the middle of the 1990s, early research into nanoscale properties and the promise of useful new products had started to attract significant funding. Many scientists were able to align their research programmes to take advantage of these opportunities, and the applications that began to emerge as progressions from previous technologies were certainly more attractive to the funding bodies than the radical ideas of molecular manufacturing with its association with 'grey goo' and similar doomsday scenarios. In this climate, many scientists preferred to distance their own research from Drexler's ideas. Then, somewhat unusually for arguments within science, the debate between the radical visionaries and the sceptics took a very public turn.

In 2001, a special issue of *Scientific American* focused on nanotechnology and contained contributions from both sides of the divide. Drexler's article was entitled 'Machine-phase nanotechnology'. In this piece, he again wrote of "bottom-up construction, in which molecular machines assemble molecular building blocks to form products, including new molecular machines". The sceptical viewpoint was argued by two highly respected experimental chemists. One was George M. Whitesides, who in his article 'The once and future nanomachine' accepted that "nanoscale machines do already exist, in the form of the functional molecular components of living cells" (an idea you will be familiar with from Chapter 3), while also noting that it would be a "staggering accomplishment to mimic the simplest living cell". However, Whitesides did question whether robotic molecular manufacturing was in fact achievable, saying that "the assembler seems, from the vantage point of a chemist, to be unworkable". He strongly advocated the view that "biology and chemistry, not a mechanical engineering textbook" pointed the way forward.

The other sceptical chemist was Richard Smalley, whom you may remember from Chapter 1 as having won a Nobel Prize for his part in the discovery of fullerenes. His article 'Of chemistry, love and nanobots' had the uncompromising subtitle 'How soon will we see the nanometre-scale robots envisaged by K. Eric Drexler and other molecular nanotechnologists? The simple answer is never.' In Smalley's view, self-replication was essential if Drexler's vision of the manufacturing potential of assemblers was to be implemented. In a 'back of the envelope' calculation, he estimated that a single assembler might take millions of years to create gram quantities of a product 'atom-by-atom'. But, at least in theory, assemblers that also produced copies of themselves could produce an ever more massive and expanding 'army' of assemblers, thus manufacturing products far more quickly. However, Smalley then argued, nanobot-controlled manufacturing was not actually feasible, even in principle, because the scale imposed impossible constraints. A nanobot could not use a manipulator arm to position an individual atom at a particular spot without controlling not only this one atom, but also all the atoms surrounding the reaction site. Manipulator arms would themselves be made of atoms and there simply would not be room to accommodate all the arms necessary to exert this control when this region was only about a nanometre in each direction. Moreover, the manipulator arm would bond to the atom being manipulated, so that it would be impossible to release the atom in the required spot. Smalley memorably described the space constraint as 'fat fingers' and the bonding problem as 'sticky fingers'.

Drexler subsequently rebutted the sceptics' arguments, but was further upset by public critiques at press conferences. In 2003 he posted an open letter to Smalley on The Foresight Institute website, correcting what he saw as Smalley's 'public misinterpretation' of his work. The subsequent correspondence between the two men, consisting of two letters on each side, was published *en bloc* and with an editorial preface in *Chemical and Engineering News* (which describes itself as 'the news magazine of the chemical world'). Drexler opened the debate by stating his own credentials as an expert, including his many scientific papers and books. He complained of the 'misdirected arguments' about multiple 'fingers' being required for the manipulation of individual atoms when 'like enzymes and ribosomes, proposed assemblers neither have nor need these'. In response, Smalley began in conciliatory fashion, acknowledging Drexler's positive contributions to the field. However, the central point of this letter was to ask what an assembler did use, if it didn't use 'fingers'. Smalley agreed that something like an enzyme could carry out precise chemical reactions, but required water and nutrients to do so, and could not build anything that was not stable in water. And how, he asked will it produce steel, copper, aluminium or titanium? The alternative, a non-water-based assembler, would seem to require 'a vast area of chemistry that has eluded us for centuries'. Interestingly, Drexler's counter did not address the issues that Smalley raised. Instead, he restated his position that molecular manufacturing is a fundamentally mechanical vision, not a biological one, and that the physical principles on which it is based are sound. He then reiterated what he sees as the fundamental concepts of systems engineering. Smalley's concluding letter noted this shift in perhaps surprisingly personal tones:

> I see you have now walked out of the room where I had led you to talk about real chemistry and are now back in your mechanical world. I am sorry we have ended up like this. For a moment I thought we were making progress. … But, no, you don't get it. You are still in a pretend world where atoms go where you want because your computer program directs them to go there.

Smalley's final comments moved away from the scientific into areas of public perception. He told of essays about nanotechnology written by middle- and high-school pupils ahead of his visit to their school, and says that of the 30 top essays he read:

> nearly half assumed that self-replicating nanobots were possible and most were deeply worried about what would happen in their future as nanobots spread around the world. … You and people around you have scared our children. I don't expect you to stop, but I hope others in the chemical community will join with me in … showing our children that, while our future in the real world will be challenging and there are real risks, there will be no such monster as the self-replicating mechanical nanobot of your dreams.

This waspish debate had a largely US focus, and obvious political overtones related to funding opportunities. (In the USA, the National Nanotechnology Initiative (NNI) oversees the funding and priorities for nanoscale research. Drexler's ideas were thought to be too 'far out' to be worthy of support, with very public spats about whether a feasibility study on his proposals was warranted. On behalf of the NNI, Smalley talked of his determination to avoid letting discussions about self-replicating nanobots halt the advance of nanotechnology.) The Drexler–Smalley debate can, therefore, hardly be implied to be revealing of the field of nanotechnology as a whole. However, the episode does tell us something interesting about the processes of communication and conflict amongst scientists, and of the importance of funding opportunities and involvement in policy decision making.

In the open letter correspondence, Drexler and Smalley essentially talked past each other. Drexler approached assemblers from a theoretical point of view based on mechanical engineering. If no physical law was broken in his analysis, then, according to his criteria, molecular manufacturing is possible, at least in theory; he considered the fact that the process had not been implemented to be irrelevant to his argument. Smalley, taking a view grounded in practical chemistry, insisted that the construction of theoretical models was not enough to demonstrate the feasibility of assemblers. An attempt to realise a nanoscale assembler in the laboratory would face insurmountable difficulties, because chemical interactions would preclude the necessary control. The chemical (indeed much of the scientific) community would share Smalley's view of the need for instrumental control of nanoscale phenomena in the laboratory as part of the practice of nanoscience.

■ Why, given these apparently irreconcilable sets of criteria, do you think that Drexler initiated the open correspondence?

▨ Drexler required a platform from which he could publicly address the sceptical scientific community. It is probable that the peer review processes would have blocked his publication in the scientific journals read by that community, since the reviewers would have been drawn from it. In any case, this kind of to-and-fro discussion is not the norm for research journals, which is why the dispute was largely carried through press conferences and the pages of more 'popular' scientific magazines.

In view of the fact that most of the applications you have encountered in your study of this topic have been progressions from earlier technologies, you might find it slightly surprising that so much of this high-profile debate revolved around the completely unrealised dream of 'radical nanotechnology'. If so, you would be in the company of many researchers in the field. One such is physics professor Richard Jones, chair of the Nanotechnology Engagement Group, a think-tank set up by the UK Government in 2005 to coordinate the discussion of social and ethical issues surrounding nanotechnology. In his 2004 book *Soft Machines: nanotechnology and life*, he wrote:

> What has been the scientists' reaction to the growing fears of grey goo? There has been some fear and anger, I think; many scientists watched the controversy about genetic modification with dismay, as in their eyes a hugely valuable, as well as fascinating, technology was hobbled by inaccurate and irresponsible reporting. But mostly the reaction is blank incomprehension. At least genetic modification was actually a viable technology at the time of the controversy, while for a self-replicating nanomachine there is still a very long way to go from the page of the visionary to the laboratory or factory. To a scientist, struggling maybe to get a single molecule to stick where it is wanted on a surface, the idea of a self-replicating nano-robot is so far-fetched as to be laughable.

> (Jones, 2004)

■ Do you think the furore over grey goo does in fact arise from 'inaccurate and irresponsible reporting' of the type that Jones points to in relation to GM?

■ The media certainly did have a part to play, but can hardly be assigned all of the blame. After all, you have read the quotation from *Engines of Creation*, which long predated any media reports. Drexler himself introduced the term grey goo; it was not an invention of the media.

So where does the responsibility of scientists lie in defining future directions for nanotechnologies? It is probably not enough for them just to make accurate statements (whether positive or negative) to the media or other bodies. If they are to counter 'inaccurate and irresponsible reporting' they need to react if it occurs. Thus:

> … maybe scientists are not entirely without blame. Most scientists working in nanotechnology themselves may refrain from making extreme claims about what the science is going to deliver, but (with some notable exceptions) they have not been very quick to lower expectations. One does not have to be very cynical to link this with the very favourable climate for funding that nanoscale science and technology has been enjoying recently.

> (Jones, 2004)

Funding always looms large in modern research!

In 2004, Drexler himself pronounced that he had shied away from the notion of self-replicating assemblers, with a revised vision of 'auto production' in which nanoscale manufacturing would depend on instructions from external computers. This was not quite a U-turn, but perhaps a recognition that scenarios involving self-replicating nanobots have coloured public and scientific assessment of his work, which in turn impacts upon the funding he was able to attract. He certainly told the scientific journal *Nature* in 2004 that he wished he had never used the term grey goo.

4.2 Debates about control

The debates about how the development and deployment of nanotechnologies should be controlled are key for many stakeholders. These debates have at their heart questions about the risk–benefit balance of research and new applications: how do we ethically maximise the benefits that nanoscience has the potential to deliver, while keeping the risks if not to a minimum then at least to acceptable levels? In practice, discussions focus on the types of application that have already been delivered or are confidently expected within the foreseeable future. The economic benefits of new materials or the health benefits of improved methods of drug delivery, say, are set against the possible toxicity of nanoparticles. Far more difficult to weigh in this balance are the unforeseen risks that inevitably go with novel technologies.

■ What kinds of control can be exercised over the development of nanotechnologies and the marketing of nanotechnology products?

■ One type of control is economic: the availability of funding for research, the money that industry can raise to bring new products to the marketplace, and the purchasing power of consumers who choose whether or not to buy the products. The other main type of control is legislative: the rules and regulations that may be imposed upon, for example, individuals, workplaces, manufacturers and sellers.

■ Thinking back to the discussion of nanoparticles and nanotubes in Chapter 1, list some of the regulatory measures that you might like to see applied to their manufacture and use.

▨ Although this list is far from exhaustive, a few options you might want to consider include:

- in relation to the production of 'free' nanoparticles and nanotubes: health and safety rules for the workplace; control of emissions to the air or to ground water; control of waste disposal;
- in relation to the use of nanoparticles in sunscreens: regulations concerning substances permitted as ingredients in medicines and cosmetics; regulations for establishing the safety of new medicines and cosmetic preparations; labelling requirements.

In the UK and continental Europe, concern has centred on whether existing regulatory frameworks covering food, drugs, cosmetics, pollution, environmental protection, and health and safety in the workplace sufficiently cover nanoparticles, or whether new forms of control are required. Taking just one example to illustrate the kinds of issue that arise, consider the triggers that determine whether, and to what extent, chemicals need to be tested before marketing or use. Currently (in 2006) these triggers take no account of particle size.

■ What assumption is implicit, therefore, in the regulation that defines these triggers, and do you regard it as justified?

▨ The assumption is that the size of a particle has no bearing on its possible toxicity. Yet, as Chapters 1 and 2 have illustrated, nanoparticles may have entirely different properties to larger particles of the same chemical composition, so this assumption is unwarranted.

The Royal Society and Royal Academy of Engineering 2004 report described this as a 'regulatory gap', and recommended that chemicals in the form of nanoparticles or nanotubes be classed as new substances under the various rules governing the testing of chemicals and the restrictions on their use.

From this one example, it is clear that regulatory regimes should be systematically and frequently reviewed. Nevertheless, many governments and their advisory bodies seem to be of the opinion that regulatory frameworks can be put in place to ensure that the risks associated with current or soon-to-be-available applications are not too serious. We must recognise, however, that such opinions are based on very incomplete knowledge. In the opinion of some individuals and pressure groups, the current state of knowledge is so woefully deficient that no regulatory system could possibly deliver the assurance of safety. They argue that, under these circumstances, the precautionary principle requires a moratorium on all the development and production of new nanomaterials while experiments are undertaken to establish the risks associated with their manufacture and use. Given the huge levels of investment in nanoscience and nanotechnologies, such a pause seems highly unlikely.

4.3 Debates with the public

At the beginning of their 2004 study, the Royal Society and Royal Academy of Engineering found very low levels of awareness across Europe regarding the issues surrounding nanotechnology. A web-based survey had been carried out in the USA in 2001, in which 57% of respondents had agreed with the statement that 'human beings will greatly benefit from nanotechnology'. But, because of the self-selected nature of the group surveyed and the lack of data on their real level of understanding of the field, it is not possible to draw from this survey any conclusions about attitudes among the general public at that time. As part of their study, the Royal Society and Royal Academy of Engineering, therefore, commissioned market research into public awareness of nanotechnologies in Britain. This research is the subject of Section 4.3.1.

4.3.1 UK attitudes in 2004

The project to determine public awareness of, and attitudes to, nanotechnologies in 2004 had two strands: (a) a survey of a sample of 1005 people, representative of the adult population of the UK, and (b) two workshops each involving about 25 members drawn from what was described as a 'broad spectrum' of the general public.

(a) The survey. The results of the survey confirmed that there was low awareness of nanotechnology among the general public, with just 29% of respondents having heard of the term and only 19% able to give any kind of definition of it. The most prevalent definitions focused on the scale of the technology or on miniaturisation. Definitions tied to specific applications, such as electronics, computing or medicine, were also common. Those able to offer a definition were then asked whether they thought that nanotechnology would 'improve our way of life in the next 20 years, have no effect or make things worse'.

■ What is your reaction to the way in which the sample of people asked this final question was selected?

▨ Clearly, this question would not have been meaningful to people who had never even heard of nanotechnology, but it is interesting that those conducting the survey chose to ask this final question only of those who could *offer* some sort of definition of nanotechnology. It appears, therefore, that it was the mere fact of a definition being suggested, rather than the accuracy or otherwise of that definition, that ensured inclusion in this sample. (From the summary of the kinds of definition that were put forward, it would appear that most were not too far off the mark, if somewhat unspecific. It might be interesting to speculate about how many people who prefer to avoid GM foods might be able to offer any kind of reasonable *definition* of genetic engineering.)

Of those able to offer a definition, 68% felt that nanotechnology would 'improve our way of life in the next 20 years' and only 4% thought it would 'make things worse'.

■ Is it valid to conclude from these statistics that the majority of the general public in 2004 took a positive view of nanotechnology development?

▨ Such a conclusion would not be warranted on the basis of this survey alone. Although the survey group as a whole was chosen to be representative of the general population, only those respondents (fewer than one-fifth of the total sample) who could offer a definition of nanotechnology were asked their view on future developments. This subgroup, therefore, probably consisted of people who were much better informed than average about science and technology and could not be said to be a typical cross-section of the public.

Indeed, it was found that this subgroup had a greater preponderance of men, of younger people and of people in higher socio-economic groupings than the original 'representative' sample. This example points up an important issue about how much science the general public can reasonably be expected to understand, how their knowledge is acquired, and how this knowledge, or lack of it, might affect people's attitudes to scientific developments and applications.

(b) The workshops. Given the experience of the negative public reaction to aspects of GM once GM products became more widely distributed, the Royal Society and Royal Academy of Engineering had considerable interest in predicting how attitudes to nanotechnology might develop as the public became more aware of it and could find out more about it. In each of the two workshops, a scientist was available to provide basic information and to answer questions about the scientific basis of possible developments. Although relatively small numbers of people were involved (50 in total), the workshops did allow a more thorough exploration of participants' ideas and interpretations. The intention and number of participants were similar to those for the 'narrow-but-deep' (NBD) component of the *GM Nation?* debate, although the group selection procedures and the ways in which exercises were run were rather different.

Box 4.1 lists the main findings of these workshops, which form an interesting 'snapshot' of some of the attitudes at the time – at least among the more interested and better informed sections of the UK public.

Box 4.1 Findings of qualitative workshops, 2004

1 Aspects of nanotechnologies about which participants were concerned

- financial implications, and the balance between investments and returns in the UK;
- the impact on society, including on the divide between the developed and the developing nations, and the influence of large corporations;
- the feasibility of some proposed applications, especially those suggested for use inside human beings;
- the possibility of unforeseen and/or long-term consequences, and whether lessons about such issues had been learned from the past (e.g. from nuclear technology and GM);

- the feasibility of control measures, especially the difficulties of enforcing these internationally;

- the role of the public: whether they would be capable of contributing to the debate, and whether that contribution would in fact have any influence.

2   Aspects of nanotechnologies about which participants were positive

- the excitement of new technologies with 'untapped potential';

- the possible applications, especially those related to medicine and, though to a lesser extent, those related to cosmetics;

- the potential for useful, and perhaps more environmentally friendly, new materials;

- the hope that quality of life would be enhanced by new products and new medical advances delivered by nanotechnology;

- what the report described as 'a sense that nanotechnology was a natural technological progression and that, in the future, arguments against nanotechnology developments will appear ridiculous'.

In general, views about emerging issues will be partly informed by what people perceive as previous analogous issues.

■ Compare the applications commonly mentioned in the 'definitions' offered by those in the survey group with the applications mentioned in the workshop summaries given in Box 4.1. Is there evidence that these applications colour the views of the public?

▒ The applications most frequently mentioned in the definitions offered by survey respondents concerned information/communications technology and medicine. From lifestyle and ethical standpoints, such applications are generally viewed in a positive light, so it is not surprising that these respondents considered that nanotechnologies were also likely to improve quality of life. The workshop participants were similarly positive about nanotechnology applications involving medicine, cosmetics and new, possibly more environmentally friendly, materials. However, there was also a harking back to more familiar technologies, such as nuclear power generation and GM; it was chiefly by association with these different technologies that some aspects of nanotechnology gave rise to concerns among the participants.

Overall, the workshops were seen as providing useful signals about ways in which a wider public debate about nanotechnologies could be fostered in the future, in relation to both the science and the stakeholders. It was clear from some of the views expressed by participants that there is unease among the general public over what is seen as 'scientists messing with the natural building blocks of matter'. Research has shown that issues such as radioactive waste and accidents at nuclear power stations have contributed to public perceptions of the risks associated with atomic-level manipulation of matter; there are similar concerns over aspects of biotechnology. Risk perception is always increased by uncertainty (although uncertainty can also fuel excitement and promise), and the supply of reliable, independent and *understandable* information is therefore essential in weighing up where the balance might lie between risks and benefits. In making judgements about the risk–benefit balance of new technologies, people also want to raise social and ethical questions. As the summary in Box 4.1 shows, these questions do not mean that the public necessarily lacks scientific knowledge (though some would argue that more detailed understanding of science among non-scientists is always desirable!), nor that it is 'anti' science or opposed to new technologies *per se*. Most often, the questions are prompted by concerns about governance: in a novel and fast-developing area like nanotechnology, who can be trusted to ensure that developments will be beneficial to society at large, who will monitor the potential risks, and how will the field be regulated? The public also wants its own voice to be listened to in any discussions about the future direction of nanotechnologies. In fact, arguments for the greater democratisation of science cut both ways. The public and some stakeholders such as environmental groups and NGOs want a say in decision making about the kind of futures that might be created from scientific advances, but other stakeholders may also have something to gain from the process; it has been pointed out that:

> In the field of environmental risk, non-technical assessments and knowledge have been shown to provide useful commentary on the validity or otherwise of the assumptions made in expert assessments … Where high levels of uncertainty exist, there may be particular benefits to opening up the risk characterisation process to a wide range of differing perspectives. (Royal Society and Royal Academy of Engineering, 2004)

From what you have discovered so far in your study of this topic, it should be apparent that the social and ethical dimensions of nanotechnological futures will certainly involve such a range of perspectives. There was a recommendation from the Better Regulation Taskforce in 2003 that the UK Government should seek to involve the public in the decision making about nanotechnology. A similar strategy was set out in a communication *Towards a European Strategy for Nanotechnology*, which stated that:

> It is in the common interest to adopt a proactive stance and fully integrate societal considerations into the R&D [research and development] process, exploring its benefits, risks and deeper implications for society. … this needs to be carried out as early as possible and not simply expecting acceptance post-facto. In this respect, the complex and invisible nature of nanotechnology presents a challenge for science and risk communicators. (European Commission, 2004)

After detailed analysis of what had been discovered during the surveys and workshop, and in full cognisance of the history of stakeholder and public dialogues conducted in the UK around both the BSE episode and the *GM Nation?* exercise, the Royal Society and Royal Academy of Engineering 2004 report broadly endorsed these positions, calling for the UK Government to initiate and fund 'a constructive and proactive debate' about the future of nanotechnology – at a stage when it could still inform key decisions and before the appearance of 'deeply entrenched or polarised positions'. In its response to the report, which was published early in 2005, the government accepted this recommendation. However, the first UK steps along this road were not taken by government, but involved a **citizen jury** sponsored by a number of very different bodies, as you will see in Section 4.3.2. Like a trial jury, a citizen jury is a panel of people, typically between 12 and 24, drawn more or less at random from the general population to address a particular issue. The jury hears and questions witnesses, then decides on its conclusions and issues a report. Fair play during the process is ensured by professional facilitators.

4.3.2 The NanoJury UK

The NanoJury UK deliberation was conducted over the summer of 2005. The project was a complex one, with the following aims:

1   To provide a mechanism for informed public views on nanotechnology to impact on policy.

2   To 'facilitate a mutually educative dialogue between people with diverse perspectives and interests, including critical and constructive scrutiny of the hopes and aspirations of those working in the nanotech-related sectors by a wider group of citizens'.

3   To explore a different kind of process by which discussion of policies for nanotechnology R&D might be broadened, with respect to both the range of issues considered and the sectors of the public involved.

The project was jointly sponsored by:

- the Interdisciplinary Research Collaboration in Nanotechnology, which is based at Cambridge University;

- the Policy, Ethics and Life Sciences Research Centre, which is based at the University of Newcastle;

- the *Guardian* newspaper;

- Greenpeace.

The oversight panel for the project also involved a very broad range of stakeholders, including academics, and representatives from the sponsors, the Research Councils, industry, and the environmental groups ETC and Green Alliance. A representative of the UK Government also sat on the panel and made a commitment that the jury's report would be considered by the appropriate governmental committee, although with the caveat that 'the results of this kind of exercise will not by themselves directly determine policy, but will provide social intelligence on the wider environment in which policy is made'. The oversight panel had the responsibility of ensuring that 'no single perspective or policy option was overrepresented' to the jury.

A science advisory panel was charged with selecting and briefing the expert witnesses who appeared before the jury, such that a wide-ranging but balanced range of views (including those from sceptics as well as proponents of nanotechnology) could be put forward, while ensuring that the technical information provided by all the witnesses was authoritative.

The whole process was observed and evaluated by an independent assessor – in fact the same one who evaluated the *GM Nation?* project.

The jury itself was set up in Halifax, West Yorkshire. Applications to join were sent to people whose names were randomly selected from the electoral roll. From those who did then apply, 20 were chosen so as to form a jury that was representative of the diversity within the local community. The jurors committed to two evening sessions a week for 10 weeks, each session lasting 2½ hours. In the first part of the process, the jury chose a topic to discuss (they decided to debate youth crime) and spent 10 sessions on that, the idea being that this would make them familiar with the process and help them to find the best way of working as a group. The remaining 10 sessions (i.e. 25 hours) were devoted to nanotechnologies. The jury was aided by professional facilitators. Typically, the witnesses gave an initial presentation. The jurors retired to formulate a set of questions, which the witnesses and a member of the science panel then attempted to answer. At the end of the process the jury produced a report and a list of recommendations.

The jurors launched their recommendations at a public event in London. Of the 20 recommendations in their final report, they highlighted four as being those that the whole jury felt most strongly about:

1   *Health*. Nano-enabled drugs could potentially considerably reduce the length of hospital stays. Improved funding mechanisms for their development should be put in place and such drugs should be available to all through the NHS.

2   *Governance*. Nanotechnologies that could bring jobs to the UK should be supported by the government in terms of education, training and research funding.

3   *Communication*. Scientists needed to communicate better. Some members of the jury had felt patronised by some of the scientists, and they had not liked the vocabulary used by the scientists. It had been noted that the scientists were not always in agreement with one another.

4   *Labelling*. Straightforward labelling in English should be applied to products that contained manufactured nanoparticles.

In general, jury members were positive about nanotechnologies, provided that safety could be guaranteed. They regarded medicine and renewable energy as the most exciting applications. They were also very positive about the jury process itself, while clearly aware of the central problem with trying to engage the public in discussions about emerging technologies; as one juror put it, 'some of this stuff is so far ahead that even the scientists aren't sure where it is going'.

Question 4.1

Comment on the conclusions of the jury in terms of the four course themes – science communication, risk, ethical issues and decision making, writing a sentence or two relating to each theme.

Summary of Chapter 4

Eric Drexler and Richard Smalley had very public discussions about the feasibility of constructing molecular assemblers. Drexler insisted that assemblers were theoretically possible from a mechanical point of view. Smalley was equally insistent that such devices could never in fact be constructed because it would be impossible to have the necessary control over the chemical interactions. This debate is illustrative of how scientists in different disciplines can sometimes talk at cross-purposes. Regardless of the actual arguments advanced, if the participants in a debate do not conduct it on the basis of shared criteria, then they are unlikely to come to any agreement.

Debates about the control of current and future applications of nanotechnology have centred on whether regulatory frameworks can be adapted to cover nanoparticles whose properties may differ substantially from those of the same substance in bulk form. In some quarters, there are major concerns about whether any regulatory framework can adequately deal with the uncertainties about how nanotechnologies might develop in the future.

Determined efforts are being made, at least in the UK, to make the public more aware of nanotechnology and to find out about general attitudes to possible future developments, so as to inform decision making. The NanoJury UK exercise was a pioneering attempt to engage previously uncommitted members of the general public in becoming informed and, hence, to give them a voice that could impact on the future direction of research, development and governmental policy.

Question for Chapter 4

Question 4.2

The NanoJury UK exercise obviously differed hugely in scale from the *GM Nation?* debate. In what other ways did the two debates differ? Did they have any similarities? In your opinion, did the NanoJury fulfil the call made in the Royal Society and Royal Academy of Engineering report for a 'constructive and proactive debate…at a stage when it can inform key decisions and before deeply entrenched or polarised positions appear'?

You may wish to look back at the discussion of the *GM Nation?* debate in Chapter 5 of Topic 6, but if you do not have the book to hand the following list highlights some of the salient aspects of that exercise:

- The main aim was to allow the exchange of information and views between experts, policy makers and the public.

- 'Stimulus' material was provided, the aim of which was to give a representative range of opinions presented in 'for' and 'against' format, rather than to supply scientific information.

- The debate took place on a very short time-scale at three different levels: six national meetings, about 40 meetings held by regional authorities and many hundreds of 'grassroots meetings'. Those who attended local meetings were self-selected. The supply of expert speakers was necessarily limited.

- Participants were all provided with a feedback form based on a specific list of questions. Thirty-seven thousand forms were returned.

- One finding was that participants were sceptical as to whether their views would have any influence, because it took place too late, once the direction of GM research was well established and there was already enormous institutional and commercial commitment to GM products.

Learning Outcomes for Topic 7

S250's Learning Outcomes are listed in the *Course Guide* under three categories: Knowledge and understanding (Kn1–Kn6), Cognitive skills (C1–C5) and Key skills (Ky1–Ky6). Here, we outline how these overall Learning Outcomes have been treated in the context of nanotechnology.

Considerations of what it means to work at the nanoscale, how properties change at this scale and how these changed properties can be understood and exploited, relate mainly to Kn1 and Kn2. Communication of scientific information in the context of nanotechnology (Kn3) is influenced by many factors, including 'hype', funding opportunities, a desire for the public to become aware of issues of concern as they emerge, the public's interest in this wide-ranging field and the fact that there is much that is still contested amongst scientists. The assessment of risk (Kn4) is an important aspect of any new technology, and you should appreciate considerations of the risks associated with nanoparticles and applications of nanotechnology for medical science. Discussions about how and in whose interests nanotechnology should develop, as well as issues about how nanotechnology development is controlled, should help you develop your awareness of ethical issues associated with nanotechnology (Kn5). An example of scientific fraud discussed in Topic 7 is relevant to both Kn3 and Kn5. By considering the early involvement of the public in the debate about nanotechnology, you should have enhanced your understanding of the various contributions to decision-making processes (Kn6).

Having studied Topic 7 you should have improved your abilities to interpret, evaluate and synthesise information and data (C1); to recognise whether claims are based on scientific evidence and consider the ethical nature of some issues (C2); to make defensible judgements about differing views, such as the debate between Drexler and Smalley (C3); to demonstrate the contribution nanotechnology can make to many contemporary issues, such as the energy debate, the progress of medical science, or the development of electronics (C4); and to apply the knowledge you have acquired to new developments in nanotechnology as they emerge (C5).

In studying Topic 7 you will have had opportunities to develop some key skills. You have received and responded to information selected from course materials and should be able to apply this to understand information received though other media (Ky1). You should be able to interpret and appropriately present qualitative data; in Chapters 1 and 2 you have also performed a number of quantitative exercises (Ky2, Ky3). You should be able to communicate information about nanotechnology clearly and correctly to a specified audience (Ky4) and may well have been able to work with others in discussion to test and clarify your own understanding (Ky5). As always, study of this topic should have helped you to plan, monitor and develop strategies for more effective learning (Ky6).

Conclusion to the course

You have now completed your studies of the seven scientific topics selected for inclusion in S250. In the process, you have frequently had to think about the relevance to these topics of the four course themes – communication, risk, ethical issues and decision making – that have been employed to represent the broader, societal, context of these illustrative topics. Marginal icons were used in Topics 1 to 4 to help you identify useful and meaningful examples of the themes, while some of the activities invited you to consider points at which a particular theme was especially relevant and to explain why. However, even in these early stages of the course it was often apparent that the themes could not be cleanly separated from one another or from the underlying science. In effect, these early examples showed that science and its context are inextricably enmeshed. Marginal icons were therefore not used in Topics 5 to 7; instead, you were expected to treat the themes in more integrated, holistic ways. Having completed your study of these topics you should have developed some of the skills necessary to investigate further examples of the rich tapestry that is science in context.

In addition to providing a broader context for science, the themes also illustrate some common trends that run through these diverse scientific topics. Indeed, if you consider science in context from the perspective of this end of the course, a possible overarching theme suggests itself – the nature of science as a human activity. In other words, the simple term 'science' is used as shorthand for a much more complex set of activities. Science as a human activity involves scientists with a range of skills and areas of expertise who produce scientific knowledge using a range of methods, for a number of reasons, taking into account various regulations, motivations and constraints. In turn these activities require timely responses from decision makers, not least in terms of addressing risk and ethical issues. And, of course, science as a human activity will circulate in wider society with the potential to reach members of the public who, in turn, will perceive and react to this information in different ways.

Having considered how the four themes may be applied to the seven topics included in S250, we hope that you will find them useful as analytical concepts should you ever find yourself having to think seriously about other major issues that have an underlying scientific basis. For instance, at the time of writing (2006), the issue of nuclear power seems to be creeping up the political agenda as a potential solution to our ever increasing demand for sustainable sources of energy – notwithstanding widely held concerns about cost, safety, radioactive waste and possible nuclear proliferation. Doubtless other important issues will emerge during the lifetime of S250. There are also bound to be significant developments in many or all of the seven scientific topics covered in the course. If we are to avoid treating unfamiliar science-based topics and issues superficially, most of us are likely to have to come to terms with some new science and consider how this science relates to its context. Moreover, if our thoughts about such topics and issues are not to be dismissed as unrealistically naïve, it is also essential that we develop the confidence to ask – and perhaps answer – probing questions about communication, risk, ethical issues and decision making.

Bearing these thoughts in mind, it is appropriate at this stage in the course to revisit each of the themes briefly to reflect upon how it was treated in the context of the scientific topics included in S250 and/or how it might be treated in the context of unfamiliar aspects of these topics or of entirely new topics. We hope you will find this exercise to have been useful when you come to tackle the forthcoming End of Course Assessment (ECA), part of which will require you to range back across the course as a whole, drawing upon material from several topics and all four themes in order to address a particular issue or question of general applicability.

Communication

Science communication pervades S250. Indeed, the course books plus material provided on DVD and the S250 course website are all examples of science communication in action in the form of distance learning materials. Further examples of science communication in action are the many interactions that we hope you have had during the year with your tutor, other students and possibly members the S250 Course Team. These may have involved face-to-face, telephone or email conversations, collaborative participation in electronic conferences and the writing of assignments upon which your tutor will have commented. This course makes no claims to cover science communication in general. Rather, it concentrates exclusively on science communication in relation to important aspects of the seven scientific topics in the course. Even so, it has been possible only to scratch the surface of this major and rapidly developing field of study. Nevertheless, it should now be abundantly clear to you – if it wasn't before the course started – just how important credible, appropriate and effective science communication can be in the modern world and just how intriguing some of the challenges that emerge from the communication of science are.

Activity C

Allow 15 minutes

The three questions posed in the *Introduction to the course* under the heading 'Communication' were:

What examples are useful to demonstrate how science is communicated?

How is scientific information communicated among scientists, decision makers and the general public?

What evidence is there to illustrate that science communication is influential?

Having considered science communication in each of the subsequent seven science topics, you should now be able to give rather more detailed responses to these questions than those provided in the *Introduction to the course*. Now try tackling an extended version of the second question: '*How is scientific information communicated among scientists, decision makers and the general public and how effectively is this achieved?*'

Spend about 15 minutes making *brief notes* about the different ways in which science is communicated among scientists, decision makers and the general public supported by *examples* drawn from S250. Are there any instances of particularly effective science communication and/or occasions when it has not worked well?

When you have completed your notes read our comments that follow immediately below. Try to resist any temptation to read them before you have had a go at this activity yourself. In any case, bear in mind that these comments represent only one of many possible appropriate responses to this question.

Scientists have a range of options available to them when they choose to communicate research findings formally, for example through conference presentations, or via posters, reports and/or publication in a peer-reviewed journal. However, by convention the main means of communicating the results of scientific research have been through peer-reviewed scientific papers in academic journals aimed at either scientists specialising in a particular area (e.g. vets and *The Veterinary Record*) or a more general, but nevertheless scientifically literate, readership (e.g. scientists and *Nature* or *Science*). Peer review is a procedure designed to check scientific work for mistakes. Therefore work that has 'passed' peer review is regarded as more credible and reliable than work that has not. Increasingly, such papers are made available over the internet – as well as, or instead of, in print – by publishers and learned societies who support 'open access' publication, or sometimes by the authors themselves, a development that should increase the overall circulation of newly published science between scientists and other interested parties.

Although the peer-reviewed scientific paper generally works effectively as a means of communication between practising scientists, there have occasionally been significant failures. For example, publication of a paper in a widely read journal does not guarantee that sufficient note will be taken of it, as you saw in the case of arsenic contamination of water in West Bengal (Topic 3). This example illustrates that there can be no guarantee that peer-reviewed published work will be read. Given the volume of scientific work published each year it is not that surprising that this is the case. Having said this, the introduction of online electronic databases, such as those accessible through the Open Library, has helped scientists to search speedily and extensively for relevant work in a wide range of academic publications.

Of course, publication after peer review is no guarantee that the results will not prove controversial. A good example here is the case of Baliunas and Soon, and their claim to have cast doubt on the interpretation of the proxy data climate record for the past millennium (i.e. the 'hockey stick' graph) endorsed by the Intergovernmental Panel on Climate Change (Topic 5). In this and other policy-sensitive areas (e.g. GM crops), different interpretations are often drawn from different sides of the debate on the basis of scientific papers published in highly respected scientific journals that rely heavily on the peer-review process.

Occasionally, peer-reviewed papers are later withdrawn when irregularities are discovered, as in the case of those published by Hendrik Schön (Topic 7). Some would argue that this is a further example of peer review failing. Others, however, would note that the extended peer-review community – that includes the full range of practising scientists – did their job effectively in identifying the mistakes in these papers and communicating them to the journal editors so that the work could be withdrawn. In short, peer review does not end when a paper has been published.

This assumes that scientists will submit their work for peer review before disseminating their findings more publicly. The dissemination of experimental results and conclusions before they have been formally peer reviewed can result in confusion and recriminations as in the case of Árpád Pusztai (Topic 6). This example illustrates how an already contested subject, that of GM crops, can become an issue of significant scientific and public controversy over whether the science had been conducted appropriately and therefore whether the findings could be considered valid and reliable.

Drawing upon either original research papers, or review papers (in which scientists try to 'pull together' recent developments in their field for the benefit of fellow scientists), a whole range of secondary sources of information keep more general audiences informed about recent scientific developments. In this way high-profile academic journals also act as a source of credible newly published scientific information for those whose main concern is the communication of scientific developments to more general audiences (e.g. science journalists working for broadcasters, print and other media). Examples of products aimed at non-specialist audiences include: science magazines (e.g. *New Scientist*, *Scientific American*); newspapers, broadcast (TV and radio) and online news outlets; dedicated websites produced by official agencies (e.g. the Food Standards Agency), NGOs and interested 'amateurs'; exhibitions in science centres and museums; and TV and radio magazine programmes that cover science. And for those interested in learning more systematically about science, there are 'popular' science books, textbooks and multimedia materials covering just about every conceivable area. In addition to this, many organisations produce posters, leaflets, booklets, videos, web pages, etc. in order to communicate certain aspects of science to particular audiences. These may range from briefing papers on quite technical areas of science (e.g. BSE/vCJD) produced by scientists for politicians, many of whom are not trained in science but nevertheless have to make key decisions about science-related issues, to leaflets produced by a supermarket chain explaining to its customers its policies on GM food.

Of course, a great deal of science is also communicated through formal courses that range from lessons in primary schools through to post-graduate and post-doctoral training. Included in this spectrum are adult education courses provided locally and courses offered by The Open University (such as *Science in Context*), the effectiveness of which can be judged, in part, by the number of students taking and passing the course. Science is also communicated very effectively through informal channels involving enthusiastic amateurs, such as astronomical and natural history societies; in the workplace; and through science fiction, fictionalised science, documentaries and docudramas (i.e. combinations of documentary and fiction) – in books, cinema, TV and radio. Judging the effectiveness of these examples can be more difficult because of the informal nature of the communication. In effect, you do not have to register to watch a docudrama on TV and there are no tests after you have viewed the programme.

So far, in considering how science is communicated, we have touched upon communication among scientists and between scientists and the public. How about communication between scientists and decision makers (other than as members of the public)? Governments employ senior advisers on both science in general and the science related to particular policy areas (e.g. medicine, animal health, food), as well as many specialist scientific civil servants. Clearly, these practising scientists need to communicate their advice to government ministers and officials, often in the form of reports (only some of which enter the public domain reasonably quickly). But this doesn't only happen at the level of national governments. Scientists around the world are central to the work of the Intergovernmental Panel on Climate Change (Topic 5) and are a major driving force behind international policy to tackle the 'greenhouse warming' problem.

Occasionally, it is necessary to carry out a major *post hoc* investigation in a science-related area (e.g. the BSE Inquiry, Topic 1) and the results of these investigations are certainly intended to influence future decision making. The legal proceedings ensuing from the discovery of arsenic contamination of drinking water in Bangladesh (Topic 3) can also be seen as an unusual example of such an investigation. The effectiveness of these forms of communication can be judged by the influence they have on future analogous issues. In this sense, the BSE Inquiry can be seen to have influenced UK government policy on GM, where the *GM Nation?* debate (Topic 6) involved inputs from scientists (and other experts) as well as members of the public. In turn, the experience of GM has also informed recent decision making, e.g. influencing debates about the emerging area of nanotechnology. For example, in 2004 the Royal Society and the Royal Academy of Engineering produced a report that was aimed at influencing decision makers in government and elsewhere in the hope that society would not come to regret decisions that are – or are not – made in the developing field of nanotechnology (Topic 7). Not surprisingly, official documents such as those mentioned above tend to be very cogently argued and authoritative. Whether or not they have long-term effects on the relationship between scientists, decision makers and the public can largely be a matter of opinion; perhaps only history can judge.

In conclusion, scientists, decision makers and members of the public now have access to much more science communication than even the most dedicated specialist or amateur enthusiast could hope to address. As a result, we need to develop skills to help us navigate through this wealth of information, and to be able to judge the credibility of a source. The best of the secondary sources of information alluded to above are very good indeed and serve an essential role in keeping non-specialists informed about important scientific developments. On the other hand, some of these sources are less than impressive as examples of science communication and a few are so misleading as to be examples of 'pseudoscience' (Topic 4). A key aim of S250 has been to help you develop skills in separating the wheat from the chaff.

Summary of key points

- Scientists formally communicate with other scientists at conferences (posters and oral presentations), and via published academic papers. They may also produce reports for decision makers and communicate with members of the public.

- Peer review is an important process for verifying that research has been conducted effectively and reported efficiently.

- Peer review does not guarantee successful communication.

- Work that has passed peer review is published and then scrutinised by fellow scientists.

- Some scientific journals are more prestigious than others. Publishing in a prestigious journal is likely to mean that more scientists read your work. It also increases the likelihood that media professionals (e.g. science journalists) will report the work in secondary publications.

- Mass media and formal science education are important sources of scientific information for members of the public.

- Official reports, e.g. from inquiries and court cases, can have long-lasting effects on public policy in relation to science.

- Members of the public are increasingly being invited to engage in discussions with scientists, decision makers and other stakeholders (e.g. NGOs) to deliberate about the future direction of scientific research.

- Secondary publications, including those on the internet, mean that there is now much more scientific information in circulation. Skills in navigating through these resources and in judging credibility are essential.

Risk

It is difficult to conceive of a topic suitable for inclusion in a course such as S250 that did not involve risk in some significant way. Some risks are entirely natural; others are either clearly human-induced or natural risks exacerbated by human activity. Some risks may be regarded as voluntary on the part of those exposed to them; others as involuntary. Some risks are objective and can even be quantified (at least in probabilistic terms); all have a subjective component to some degree. Some risks are short-term; others are long-term. Some risks are familiar and well-understood; others arise through the exploration of new areas of science or their development in the form of novel technologies. Some risks affect relatively small numbers of people (at least at any one time); others are of sufficient magnitude to threaten the stability of human society, the survival of the human species or even the whole biosphere.

It is clear that there is no such thing as absolute safety. Instead, when assessing risk it is more useful to think in terms of relative safety. We cannot avoid making decisions about exposure to risk and risk mitigation. Since choices entail value judgements, we are immediately confronted with a range of other factors, including ethical issues, which also need to be addressed. Information (including scientific information) required for effective decision making clearly needs to be communicated to scientists, decision makers and members of the public, but what information, when, to whom and how? Similarly, both the decisions themselves and the bases upon which they have been made have to be communicated clearly to all those affected by them.

Activity R

Allow 15 minutes

The three questions posed in the *Introduction to the course* under the heading 'Risk' were:

What aspects of risk are covered in the course?

How are risks identified?

How do you assess and manage risk?

For this activity, let us effectively combine the second and third of the above questions: '*In general terms, how can the possible risks associated with a new scientific development be identified, assessed and managed?*' As well as thinking through what might be involved in such a hypothetical development, consider what has actually happened or is happening in the cases of the scientific topics covered in S250.

As in Activity C, spend no more than about 15 minutes making *brief notes* in response to this question before reading our comments immediately below. As always, many other responses would be equally appropriate.

Risks emerge in a number of different ways, including anecdotal observations, such as farmers noticing unusual behaviour in their dairy cattle (Topic 1). In an ideal world things would be more organised; risks would be identified by a thorough risk assessment of any scientific or technological development *before* it commences. However, nobody predicted BSE – let alone vCJD – before changes were introduced in cattle feed production. Neither was the water drawn from Bangladeshi tube-wells analysed for arsenic until long after serious problems had been reported in large numbers of people. Of course, careful monitoring might at least have revealed these problems at much earlier stages. But even this approach is not fail safe. Although monitoring (following on from thorough pre-release testing) is standard practice in the pharmaceuticals industry, drugs still occasionally have to be withdrawn when problems are later discovered that affect only small sub-sets of patients, that appear only when drugs are used in combination, or that become apparent only after many years of use. In these instances, where longer term risks can be unpredictable, medical professionals and patients should make judgements based on the patient's current prognosis and level of suffering, in combination with the potential for a drug to relieve symptoms (and potentially extend life) and/or reduce pain. In so doing, the patient has some level of control over their treatment, an example of informed consent.

The assessment of risk would normally be part and parcel of a risk's formal scientific identification. Even if the risk cannot be quantified (e.g. by calculating the number of people exposed to it or the probability that an individual in a particular population would be affected by it), then it may still be possible to assess it qualitatively (e.g. from remote to likely, trivial to serious). But how can the risks associated with entirely new (potentially large scale) developments such as genetic manipulation of organisms or nanotechnology be assessed?

Identifying and assessing the possible risks associated with novel products that result from either genetic manipulation or nanotechnology pose particularly difficult problems in trying to anticipate all their possible short- and long-term effects. In these instances risk assessments may be based on similar products using the concept of 'substantial equivalence'. However, this can be controversial because there can be no guarantee that novel products will react in the same way as existing ones once they are in the natural environment. One way of addressing this issue is to invite members of the public to participate in deliberative exercises to assess what level of risk they are willing to take on in terms of the potential benefits a novel product may offer. This is a form of informed consent, even if only a small part of the information is available. It is also an attempt to increase democracy in scientific decision making, potentially reducing concerns about society being exposed to involuntary risks.

In contrast, we understand clearly the problem of NEOs – if a large one were to impact the Earth, it could cause a lot of damage. Scientists (and amateur astronomers) can systematically search for NEOs and even quantify their likelihood of impact in probabilistic terms for various dates quite far into the future. As Topic 2 showed, choosing when, how and to whom to communicate the results of these risk assessments can be a much harder issue to resolve.

Identification of the risks posed by (human-induced) climate change is different again. The inadequacies of available data and the sheer complexity of the Earth's climate system have long created difficulties for scientists seeking to identify the 'signal' of a human influence on global climate, and to assess the potential consequences of allowing greenhouse gases to go on accumulating in the atmosphere. Though the work of the IPCC has done much to consolidate a broad scientific consensus on the causes and implications of recent climate change, there are still many uncertainties about how the effects of 'greenhouse warming' will evolve over the decades ahead. This uncomfortable reality is a central part of the climate change challenge.

Assuming that risks have been identified and assessed in some way, there is a responsibility to manage these risks. Clearly such management involves decision making, in which ethical considerations (and other factors, such as economic considerations) inevitably play a part and in which communication is very important. Other than monitoring NEOs and, if necessary, making contingency arrangements, there is little that society can do at present to manage the risks posed by NEOs. However, at some stage in the future humankind may decide to try to deflect a potentially hazardous NEO, given sufficient notice.

Although the track record so far is not particularly impressive, it ought to be possible to prevent or at least minimise any adverse consequences from episodes similar in scale to those that led to BSE/vCJD or arsenic in Bangladeshi well water. The findings from the BSE Inquiry (Topic 1) and the court case (Topic 3) should help to inform future risk management in these areas. The novelty associated with the genetic manipulation of organisms and nanotechnology (Topics 6 and 7) suggests that extreme vigilance is necessary and that swift action would be required if and when any unexpected risks were detected. The uncertainties that remain to be resolved in the science of climate change, and the highly complex socioeconomic, political and ethical issues that surround efforts to address this global problem (Topic 5), show just how difficult the management of risk is likely to be in situations other than the simplest and most straightforward. However, it is possible for members of the public to make a difference by taking their own risk management measures. In terms of climate change, options include efforts to reduce your own energy consumption (e.g. switch to low-energy bulbs, energy-efficient domestic goods, home insulation, fuel-efficient cars, etc.) or investment in your own 'renewable' energy systems (e.g. solar water heaters or solar photovoltaic panels on the roof).

Summary of key points

- Risks can be identified in a number of different ways. For example, anecdotal observations from vigilant members of the public were combined with scientific risk assessments to identify arsenic in well water in Bangladesh.

- Quantitative and qualitative assessments of risk can help to inform how the risks should be managed.

- Informed consent has the potential to shift an involuntary risk to a voluntary one. Exercises that involve the public in decision making about novel areas of scientific investigation may have the same effect.

- Knowing when, how and where to communicate risk to scientists, decision makers and members of the public is not an easy issue to resolve, particularly when there is disagreement about how best to manage those risks (e.g. NEOs, climate change) and unforeseen risks.

Ethical issues

Ethics is an important branch of the academic discipline of philosophy. It would have been far too ambitious for S250 to have attempted to tackle ethics *per se*. Instead, we have restricted our scope to what we have called 'ethical issues', by which we mean consideration of the question 'What should happen?' in circumstances in which the answer is not clear-cut. We have therefore mainly avoided trying to provide answers (except in a few cases where individual authors have given their *personal* views by way of illustration).

Much of modern science is extremely expensive. It is funded by a variety of sources, including the public purse, industry and charities. Since public funds are always limited, society must decide how much to spend on science as opposed to education, health, transport, defence, development aid, etc. It therefore seems appropriate that some combination of wider society, decision makers and the science community should decide how much of the science budget to allocate to different areas of science and so on down to the individual researcher who must decide how best to deploy the staff, time, equipment and materials available to her or him. So, one ethical issue that almost always pertains is how limited resources should be allocated in science-related activities, i.e. what areas of science are worthy of public funding? Of course, there is an equivalent ethical debate in terms of funding by charities whose resources are often much smaller in comparison to the public purse. Private funding takes this further still because, although greater funds may be available, a profit motive may be involved. Decisions about which research projects to fund will therefore involve some assessment of the likely financial return on the investment.

By definition, 'frontier' science – and particularly the technological developments associated with frontier science, such as GM crops – involves stepping into the unknown and this may entail unforeseen risks. The potential benefits of these developments, such as greater crop yield and reduced use of pesticides, must be balanced against these possible risks – some of which can genuinely be unpredictable. Should this be used as a reason to restrict those scientific developments that are allowed and those that are not? And then there is the question of who (and how many) might benefit from such developments or be exposed to possible unanticipated risks from them and whether their exposure is voluntary or involuntary. In other words, an ethical judgement could be made based on how many might benefit against the possible risks. (An ethical judgement of this nature could be based on the utilitarian premise that the needs of the few are outweighed by the needs of the many.) A second ethical issue is therefore whether actual or potential risks are deemed to so outweigh potential benefits that certain areas of science should be controlled so tightly that this might inhibit or prevent their development. The challenge, of course, is to decide who should make a decision of this nature, when and how?

Furthermore, even if society were to decide that it could afford to support the development of an area of science and that it considered the potential benefits of development to outweigh any potential risks, there might still be a question of whether it is 'right' in some sort of ethical or moral sense to allow development. While supposedly 'objective' cost-benefit analysis might assist society in taking rational decisions with respect to resources and risks, such an approach cannot provide an infallible guide as to what we 'should' or 'should not' do when it comes to the ethical or moral sphere. Some people would support 'the greatest good for the greatest number' type arguments and take a 'utilitarian' stance on many or most issues; others would take (or try to take) an 'absolutist' position on at least some issues, for example by rejecting any form of animal experimentation regardless of the potential benefits. Although deeply-held opinions concerning ethics or morals are unlikely to be revised readily, keeping open channels of communication between those with different ethical views may help to maintain a level of social cohesion.

Activity E

Allow 15 minutes

The three questions posed in the *Introduction to the course* under the heading 'Ethical issues' were:

Which ethical issues will be covered in this course?

How might the purposes of scientific investigation be judged as an ethical issue?

How might the processes of scientific investigation be judged as an ethical issue?

For this activity, we would like you to think about the issues raised by the second and third questions above by tackling the following question: '*When it comes to the distinction between the purposes and processes of scientific investigations, what are the responsibilities of practising scientists and of society more generally?*' Illustrate your response by drawing upon examples from S250 or elsewhere.

As in previous activities, spend no more than about 15 minutes making *brief notes* in response to this question before reading our comments immediately below. As always, many other responses would be equally appropriate.

One important point is to require that in planning and undertaking their work, scientists behave honestly. For example, they should be open about their motives and publish results that genuinely reflect their findings. You'll appreciate from the work of Hendrik Schön discussed in Topic 7 that this is not always the case. Indeed, some commentators are concerned that well-publicised cases of this type are just the 'tip of the iceberg'. However, the fact that such cases come to light relatively rarely, when other scientists are unable to reproduce the published work of others, suggests that the great majority of scientists do pursue their work responsibly and with integrity.

How do scientists decide what work they undertake and how? It is always *necessary* for practising scientists to decide for themselves whether or not the purposes (ends) of the work they propose to undertake and the approaches (means) they intend to adopt are justified. But it doesn't follow that such individual approval is *sufficient* for it to be allowed to proceed. In other words, for scientific work to be allowed to take place scientists should be required to demonstrate that it meets at least the ethical standards embodied in current legislation and adheres to relevant regulatory frameworks (e.g. by submitting plans for any proposed work for consideration by ethical committees).

Very few scientists – at least these days – have sufficient financial independence to fund 'cutting edge' scientific research without recourse to the public purse, or other sources of funding, such as charities or industry. When scientists are given access to such funding they cannot be allowed to decide upon the direction and nature of their research entirely without reference to the views of those providing the funding, or to wider society. There should be some accountability. Apart from any other consideration, resources spent on one project are not available for another. However, even if funding issues could be set aside, neither scientists nor anyone else can be regarded as morally free agents. Everything we do – including the pursuance of scientific developments – we do as members of human society. The outcomes of our actions have the potential to impact on others, but also on the rest of the biosphere as well, sometimes with long-term consequences. Thus, it could be argued that everyone ultimately has *some* responsibility for the consequences of all scientific developments, however funded and whether carried out in their name or not. Does it follow, therefore, that those (potentially) affected should have the right to veto a research initiative? If so, how might this be achieved? Is a simple working majority in favour of a research initiative sufficient to consider it to be ethically acceptable?

In an ideal world, wider human society should satisfy itself in some way that both the purposes and the processes of *all* scientific developments are justified. If it is not satisfied about either or both of these criteria, then it has an obligation to do its best to change or prevent these developments. Of course, this is easier said than done. For a start, the world is divided into many more-or-less independent legal jurisdictions that produce legislation and regulatory frameworks according to their political system, cultural traditions, etc. If a particular country decides to allow or even encourage its scientists to pursue particular scientific developments, then even widespread disapproval of either the purposes or processes of these developments by citizens of other countries is likely to have only a limited effect (certainly no more so than their influence in other areas such as human rights). However, public pronouncements by high-profile world leaders can be influential, as can international treaties, direct diplomatic intervention and/or international trade sanctions. Some might even argue for military intervention, although this is unlikely to be pursued with respect to *scientific* research (even if it might be considered with respect to certain *technological* developments).

How about the situation *within* a particular country (and hence legal system and regulatory framework), when proposed scientific developments do not command unanimous approval with respect to purpose and/or process? Suppose that individuals genuinely disapprove of either the ends or the means of a particular scientific development. Their first recourse would presumably be to use any legal devices open to them to have the development outlawed or at least not funded. This may include public protests to draw attention to – and gain support for – arguments that they may be putting to political representatives (or those who aspire to represent us). What if this approach doesn't work – because the majority of the population is either indifferent to the case being made or positively approves of the proposed development? It is easy to see that it is but a small step from this point to potentially serious conflict with the human rights of other people (for instance, to engage in perfectly legal and tightly controlled – but evidently controversial – scientific research or development). There are no easy answers when ethical issues are in such direct opposition.

These issues come to the fore particularly starkly in those cases in which research into diseases or their treatment entail experiments or tests on animals. Laudable though it might be to work towards curing a major human disease (purpose), some people would draw a line before *any* experimentation on humans or sentient non-human subjects (processes). Others might argue that no matter what the potential benefits offered by genetic manipulation or nanotechnology, the potentials risks are sufficiently great that further developments ought to be either prevented or severely restricted.

Summary of key points

- Scientists have a responsibility to take account of ethical issues (in particular, the purposes and processes) related to their work, working within relevant legal and regulatory frameworks that govern scientific research.
- Those likely to be affected either directly (e.g. a patient considering participation in a drug trial) or indirectly (e.g. a patient's relatives and friends) by scientific research should think seriously about both the purposes and processes of any proposed scientific developments.

- Scientific developments may have long-term effects and it is important that those considering ethical issues address this possibility (e.g. climate change is likely to affect future generations).

- Ethical value systems are likely to vary both within and between different countries and cultures. Reaching agreement in these instances can be extremely difficult.

- Communication, through open dialogue and consultation may help to maintain social cohesion when a range of ethical views are in evidence.

Decision making

As we have seen above, scientists, decision makers and members of the public are continually faced with choices that have to be made, including:

- Whether or not to provide resources for the development of particular areas of science (or ensuing technological exploitation) and, if so, how much.

- Whether or not the possible risks of allowing development are sufficiently outweighed by the potential benefits to permit or encourage the continuation of such development. And, following the successful introduction of certain developments (e.g. new foodstuffs and medicines), whether or not to consume and/or be treated by them.

- Whether or not possible ethical or moral objections to the development of an area of science – or to the prevention of its development – should be allowed to influence decisions about the development, irrespective of issues of cost and/or safety.

- How to provide effective channels of communication between scientists, decision makers and members of the public to discuss these and other issues relevant to science in context.

Decision making in science – and in most other areas of public life – is seldom a clear-cut matter of unqualified acceptance versus complete rejection. How then can we make effective decisions about science? Ideally, multiple rounds of discussion and consultation involving all interested parties should precede important scientific and/or technological developments. These discussions should allow participants to agree on the terms of their deliberations and on the framework for the decision making process. These debates should be staggered in time (across years or even decades), allowing participants to take account of ongoing developments, and involve numerous separate decisions taken at a number of different levels (by individuals, organisations and institutions, and national and international politicians).

The reality, of course, is more complex, not least because of the challenge of knowing when to begin discussions; when is an emerging issue sufficiently mature to warrant extensive debate? One way of deciding when to begin public discussions is as soon as significant promotional activities are in evidence. In this way, decision makers can monitor attempts by stakeholders (including scientists, industry, NGOs and others) to promote their views, for example through direct lobbying with decision makers, advertising campaigns and mass media accounts. In addition, decision makers need to take account of new media as it is also clear that the internet is becoming increasingly important in empowering members of the public, across countries, to form online networks of support, discussion and activism about science-based issues. (Of course, members of the public can also contact their elected representatives via traditional means of communication, such as letters.)

Activity D

Allow 15 minutes

The three questions posed in the *Introduction to the course* under the heading 'Decision making' were:

Who are the decision makers within each of the topics covered in the course?

What measures can be taken to protect the public?

What role does the public play in decision making about science?

For this activity, we would like you to tackle an extended version of the third of the above questions. '*Should the public play a role in decision making about science and, if so, to what extent and how might this be achieved?*'

As in previous activities, spend no more than about 15 minutes making brief notes in response to this question before reading our comments immediately below. As always, many other responses would be equally appropriate.

The easiest way in which members of the public might play a role in decision making about science is simply to participate periodically in electing political representatives and then leave the decision making to them (*via* government ministers, civil servants and then indirectly through the legal and regulatory frameworks that govern science). In the UK this has been the most obvious form of democratic engagement in science-based issues since universal suffrage was granted to adult women and men. Of course, the better informed the electorate (which includes scientists, decision makers and members of the public) is about science in context, the more likely it is to bear such matters in mind when campaigning and voting. Nevertheless, it would be extremely naïve to assume that science-related issues (including enormously significant ones such as climate change, genetic manipulation or nanotechnology) would feature as more important than levels of taxation or spending on health, education, defence, etc. for more than a minority of voters. And this is likely to affect how political parties campaign on these issues. In practice, this means that important science-related issues may not feature extensively in public debates during election campaigns.

A further layer of public engagement in decision making about science is also already in evidence, involving regulatory bodies and ethics committees established to control experimental treatments of human patients and research experimentation involving animals. In addition to relevant experts (e.g. scientists, medical professionals, bioethicists, etc.), such committees usually also include lay members. The power of these lay representatives to influence expert decision making in relation to the business of these committees and bodies is, of course, difficult to establish. However, it would be hard to imagine a situation in which lay representatives were given the opportunity to lead the agenda for decision making in these fora.

More recently, high profile emerging science-based issues, such as GM (through *GM Nation?* – Topic 6) and nanotechnology (through NanoJury – Topic 7), have been subjected to systematic attempts to engage members of the public in decision making about these issues. In part, this is a response to concerns raised by the BSE/vCJD episode (Topic 1). Could it now reasonably be argued that the public has *no* role to play in decision making about science? Not only are many scientific developments financed, at least in part – directly or indirectly – through the taxes we pay, members of the public may be exposed to any ensuing risks (as well as be recipients of any possible benefits). There is clearly a lot to be learnt from such exercises, both about the issues debated and about how scientists, decision makers and members of the public can engage effectively with each other on such complex issues. However, it is probably unlikely that members of the public, scientists or official decision makers would retain much enthusiasm for such large-scale consultation exercises if they became routine for all scientific research. The vast cost of such exercises in terms of time and resources would also be likely to detract from other publicly-funded activities, again making this an unattractive option. The question remains, of course, as to how and why certain areas are chosen for deliberation over others.

Summary of key points

- Decision makers need to take account of risk and ethical issues (and relevant other factors), and decide when and how to communicate these issues to scientists and members of the public.

- Scientists and scientific institutions, industry representatives and NGOs regularly promote their views to other decision makers (including members of the public) through lobbying, advertising campaigns and mass media accounts.

- Electorates in the UK and elsewhere have opportunities to influence science policy by voting periodically in local, general and sometimes transnational elections. However, science has to compete for space against high-profile issues such as health, education, etc. As such, science-related issues may not feature strongly in election campaigns.

- Lay members are regularly included on ethical committees and regulatory bodies. However, their power to lead the agendas of these bodies may be limited.

- Exercises in public engagement and deliberation have introduced opportunities for dialogue and engagement between scientists, decision makers and members of the public. Choosing which issues to discuss, when and how to conduct these exercises, and who to involve is, in itself, subject to considerable debate.

- Ultimately, members of the public may choose to avoid novel products, such as foodstuffs and medicines, because of perceived/assessed risks, on ethical grounds or for other reasons.

Epilogue

New science-related issues emerge (or old ones re-emerge) every now and then. For instance (as noted briefly above), the possibility of rejuvenating the nuclear power industry, after several decades in which this subject received hardly any mainstream discussion, is currently (2006) being discussed by prominent politicians and others. As a Course Team we are confident that other science-related issues will also emerge (or re-emerge) as the course progresses through its presentations.

As a result of studying S250 *Science in Context* we hope that you will have developed the knowledge and skills necessary to cope with *both* the science underpinning the seven topics (and possibly other topics) *and* science's broader societal context (as exemplified by the four course themes). You should thus be better equipped to engage effectively in discussions about science-based issues as a scientifically informed citizen. We also hope that S250's emphasis on Learning Outcomes will have encouraged you to reflect on *how* you learn and thus become a more effective life-long learner.

Answers to questions

Question 1.1

(a) Converting all the diameters to metres and rounding at this stage to just one significant figure (which is acceptable because the final answer is required to within only one order of magnitude):

$$D_{Earth} \quad = 6371 \text{ km} = 6.371 \times 10^3 \text{ km} = 6.371 \times 10^3 \times 10^3 \text{ m} \approx 6 \times 10^6 \text{ m}$$

$$D_{tennis\ ball} = 6.5 \text{ cm} = 6.5 \times 10^{-2} \text{ m} \approx 7 \times 10^{-2} \text{ m}$$

$$D_{buckyball} = 0.7 \text{ nm} = 7 \times 10^{-1} \times 10^{-9} \text{ m} = 7 \times 10^{-10} \text{ m}$$

Therefore $\dfrac{D_{Earth}}{D_{tennis\ ball}} = \dfrac{6 \times 10^6 \text{ m}}{7 \times 10^{-2} \text{ m}} \approx 10^8$ to the nearest order of magnitude

and $\dfrac{D_{tennis\ ball}}{D_{buckyball}} = \dfrac{7 \times 10^{-2} \text{ m}}{7 \times 10^{-10} \text{ m}} = 10^8$ to the nearest order of magnitude

Comment. Put the other way round, if you could scale up a C_{60} buckyball until it was the size of a tennis ball, then the same scaling factor would inflate the tennis ball until it was the size of the Earth!

(b) $1 \text{ day} = 1 \text{ day} \times 24 \dfrac{\text{hours}}{\text{day}} \times 60 \dfrac{\text{minutes}}{\text{hour}} \times 60 \dfrac{\text{seconds}}{\text{minute}} = 86\ 400 \text{ s}$

Fingernail growth $= 0.1 \dfrac{\text{mm}}{\text{day}} = \dfrac{1 \times 10^{-4} \text{ m}}{86\ 400 \text{ s}}$

$= 1.157 \times 10^{-9} \text{ m s}^{-1}$

$\approx 1 \text{ nm s}^{-1}$ to one significant figure

A nanometre is roughly how much one fingernail grows in 1 second.

Question 1.2

You may have considered the following:

In relation to hazard, studies would seek to quantify the 'potential to cause harm' if nanotubes were inhaled. Tests could be undertaken to establish important physical properties of particular types of nanotube, such as:

- Are their aerodynamic properties such that they could float in the air and hence be easily inhaled (e.g. by those involved in research or manufacture)?
- Are they of suitable dimensions to reach the deep lung?
- Are they too long to be easily removed by macrophages?

Toxicology studies – probably on small mammals (which mean that the studies themselves raise ethical issues) – would be required to find out what happens once tubes actually enter the lungs. For instance:

- How robust are the tubes within the lungs?
- Do tubes cause a reaction, such as irritation, inflammation or scarring?
- Does long-term exposure cause cancer?

In relation to exposure, questions to be answered would include:

- Do research or production methods result in the release of nanotubes in such a way that workplace staff could be exposed to significant concentrations in the air? (Of course this begs the question 'What concentrations should be regarded as significant?', which is linked to the toxicology data.)
- What procedures are in place for routinely controlling airborne emissions and for dealing with accidental release?
- Could members of the general public be exposed to nanotubes in the air?

This list is far from exhaustive!

Question 1.3

See Figure 1.25a and b.

Figure 1.25 Answer to Question 1.3. (a) Constant-height mode; (b) constant-current mode.

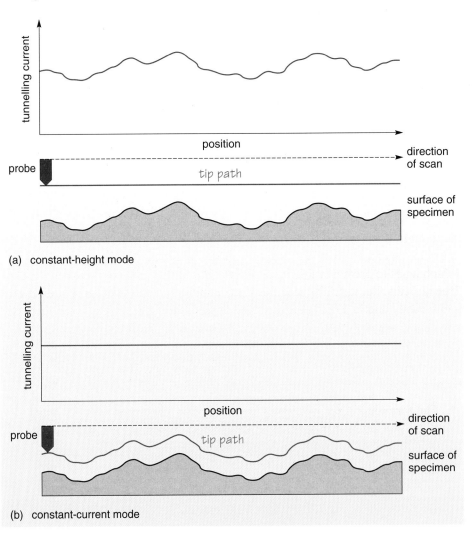

(a) constant-height mode

(b) constant-current mode

Question 1.4

(a) Each face of the cube has a surface area of $(10 \text{ nm})^2 = 100 \text{ nm}^2$.

There are six faces (top, bottom, front, back, and two sides), so the total surface area of the cube is 600 nm^2.

The volume of the cube is $(10 \text{ nm})^3 = 1000 \text{ nm}^3$.

The surface-to-volume ratio for the cube is $\dfrac{600 \text{ nm}^2}{1000 \text{ nm}^3} = 0.6 \text{ nm}^{-1}$

(b) For a sphere to have the same volume as the cube, it must be of radius R, such that

$$\frac{4}{3}\pi R^3 = 1000 \text{ nm}^3$$

So $R = \sqrt[3]{\dfrac{3 \times 1000 \text{ nm}^3}{4\pi}} = 6.2 \text{ nm}$

(c) The surface area of that sphere is $4\pi(6.2 \text{ nm})^2 = 483 \text{ nm}^2$.

The surface-to-volume ratio for the sphere is

$\dfrac{483 \text{ nm}^2}{1000 \text{ nm}^3} = 0.5 \text{ nm}^{-1}$

(to one significant figure).

For a given volume of material, the cube has the higher surface area, so would be a better choice of shape for an application where this was important, as it is in heterogeneous catalysis.

Question 1.5

Two main factors may mean that nanoparticles pose greater health risks than larger particles of the same composition. First, for the same mass of material, the nanoparticles will have a greater surface area than the larger particles, which leads to greater reactivity and to the possibility of the generation of more free radicals. Second, it is probable that nanoparticles are taken up by cells both more easily and in a different way than larger particles.

Question 1.6

(a) A sketch diagram of the first few generations of this dendrimer is shown in Figure 1.26.

(b) The formula $Z = n \times b^G$ applies equally to this dendrimer, with $n = 4$ and $b = 2$. (If you are unsure about this you can test it against your sketch.)

For $G = 5$, $Z = 4 \times 2^5 = 4 \times 32 = 128$

Figure 1.26. The first few generations of a dendrimer based on an ethylenediamine core (schematic representation).

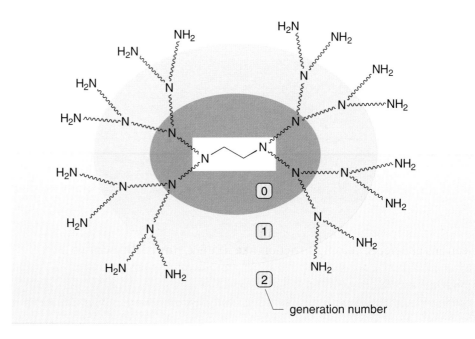

Question 1.7

New nanotechnologies are indeed likely to lead to smaller components for use in electronics (and in other applications). However, nanotechnologies and miniaturisation are far from being equivalent concepts. Nanotechnologies seek to exploit novel structures and properties of matter at the nanoscale, which are significantly different from those of bulk materials.

Question 2.1

A cube in which 10 atoms fitted along each side would have $(10^3) = 1000$ atoms in total.

In a similar way, since the average quantum dot has about 50 atoms across a diameter, i.e. 25 atoms along a radius, the total number of atoms in a dot is

$\frac{4}{3}\pi \times 25^3 = 6.545 \times 10^4$ (to 4 significant figures)

We know that 6.02×10^{23} atoms have a mass of 28 g, so there are

$\frac{6.02 \times 10^{23} \text{ atoms}}{28\text{g}} = 2.15 \times 10^{22} \text{ atoms g}^{-1}$

and $\frac{2.15 \times 10^{22} \text{ atoms g}^{-1}}{6.545 \times 10^4 \text{ atoms/dot}} = 3.285 \times 10^{17} \text{ dots g}^{-1}$ (to 4 significant figures)

To avoid rounding errors, 4 significant figures have been carried through the calculation, but this answer should be rounded to no more than 2 significant figures, which is all that can be justified given the precision of the data.

The final estimate for the number of silicon quantum dots is therefore

3.3×10^{17} dots g^{-1}

Question 2.2

The resultant has magnitude that can be calculated by Pythagoras, or be measured, to be approximately 5 km at an angle θ of approximately 51° (Figure 2.23).

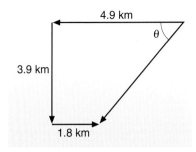

Figure 2.23 Scale diagram of Dr Vector's route to work.

Question 2.3

You need to go four steps in the direction of $\hat{a}_1$ and then two steps in the direction of $\hat{a}_2$ as shown in Figure 2.24, so $(n, m) = (4, 2)$.

Question 2.4

In a metal the electrons in the conduction band conduct electricity as in the free-electron model. Electrical resistance is caused by the scattering of electrons at defects, causing a friction-like heating. Any heating causes the atoms in the lattice to vibrate more, leading to more collisions and more scattering, thus increasing the resistance. In a semiconductor the current is carried by electrons that are thermally excited from the valence band into the conduction band and by the holes left behind. As the temperature increases, more electrons are excited into the conduction band, the number of charge carriers increases and so the conductivity also increases.

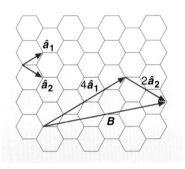

Figure 2.24 Illustration of how the values of n and m are used to construct the chiral vector **B**.

Question 2.5

(a) stress $\sigma = \dfrac{F}{A}$.

Force is the weight $W = \text{load} \times g = 6.0 \times 10^3 \text{ kg} \times 9.8 \text{ m s}^{-2} = 5.880 \times 10^4 \text{ N}$

The area A of a circle of radius 7.5 cm is

$A = \pi r^2 = 3.14 \times (0.075 \text{ m})^2 = 0.0177 \text{ m}^2$

So

$$\sigma = \frac{F}{A}$$
$$= \frac{5.880 \times 10^4 \text{ N}}{0.0177 \text{ m}^2}$$
$$= 3.32 \times 10^6 \text{ N m}^{-2}$$
$$= 3.3 \text{ MPa (to 2 significant figures, which is the precision of the data given)}$$

(b) strain $e = \dfrac{\sigma}{E}$, where σ is the stress and E is the value of Young's modulus.

So here

$$e = \frac{\sigma}{E}$$
$$= \frac{3.32 \times 10^6 \text{ Pa}}{2.0 \times 10^{11} \text{ Pa}}$$
$$= 1.66 \times 10^{-5}$$
$$= 1.7 \times 10^{-5} \text{ (to 2 significant figures)}$$

(c) change in length $\Delta L = eL$.

Here that gives

$$\Delta L = eL$$
$$= (1.66 \times 10^{-5}) \times 3.0 \text{ m}$$
$$= 4.98 \times 10^{-5} \text{ m}$$
$$= 5.0 \times 10^{-5} \text{ m (to 2 significant figures)}$$

Question 2.6

It is not clear what the fate of nanoparticles might be once released into the environment, i.e. where they might end up. The concern is because relatively little is understood about the effects of the high surface reactivity of nanoparticles on organisms and ecosystem processes. The two main concerns then relate to nanoparticles' fate and toxicity.

Question 3.1

Ribosomes may come to mind as assembly lines that manufacture different types of protein. You might know that a number of ribosomes are able to attach to an individual mRNA molecule, so that a succession of identical protein products is formed, increasing the capacity of the protein assembly line. DNA is another good example, able to self-replicate by the process described in Section 2.1 of Topic 6. Proteins themselves are powerful biological nanomachines. Enzymes, for example, retain their capacity to bring about complex chemical transformations accurately and repetitively even when they are removed from living systems. One example is the enzyme RNA polymerase that you know of from Topic 6, which is able to move along thousands of DNA base pairs without becoming detached. Finally, the mitochondrion (the cell's main energy generator) is another proficient nanomachine, generating ATP. This was referred to by Sir Harry Kroto in the movie sequence, where he enthused about 'these spinning electric motors, made by biological systems millions of years before Faraday'. We will look at the function of these 'electric motors' later in this section.

Question 3.2

With peptide nanotubes, there are claimed advantages with respect to speed of action and lower risks of development of resistance. With Desai's insulin-releasing capsules, a major advantage is the avoidance of the need for immunosuppression after transplantation.

Bear in mind, however, that with each of these new technologies there may be unforeseen risks that have to be taken into account in assessing what types of treatment are 'best'. Keep the question of advantages in mind as you read about the other nanotechnological techniques in medicine mentioned in the rest of this chapter.

Question 3.3

Yes; two examples are mentioned. Self-assembly is important in the construction of ribosomes, which contains different types of protein and RNA. Proteins also become associated with lipids by self-assembly, as with a variety of membrane-bound complexes, such as the ATP-synthase enzyme.

Question 3.4

Rotary motors of the type evident in mitochondrial ATP-synthase and related molecules could be thought of as a type of wheel – and, furthermore, one that is used in some bacteria to achieve locomotion!

Question 3.5

By using restriction enzymes, of the type you read about in Section 3.1 of Topic 6. Their use can leave unpaired base sequences at the end of the construct and these can prove useful as points to join together DNA sequences, or even create 'hooks' for other types of attachment.

Question 3.6

Most conventional antibiotics act at particular target sites, e.g. they act on particular receptors in the membrane or at the active sites of specific enzymes. Changes in the chemical make-up at such localised sites (i.e. via mutation and the natural selection that results) will reduce the effectiveness of the antibiotic effect. The effect of peptide nanotubes appears to be more dispersed, which suggests that resistance to them would only come about with more substantial and widespread changes, which are less likely.

Question 3.7

Section 3.3.3 explained how individual dendrimers with distinct functions can be joined together via short, single-stranded pieces of DNA, which come together via complementary base pairing (Figure 3.13).

Question 3.8

The use of nanoparticles does raise health concerns, such as possibly toxicity, but the blockage of capillary vessels in this way is not one of them, for reasons to do with their relative sizes. Figure 1.2 reveals the size difference between cells and nanoparticles. Dendrimers, for example, are similar in size to proteins. A typical cell, a red blood cell for example, is about 7000 nm in diameter (see Figure 1.1) and capillaries are at least wide enough to allow the passage of such cells. Haemoglobin (contained within red blood cells) is 7.0 nm wide. The maximum dimension of ribosomes, for example, composed of many individual proteins, is about 70 nm. The largest nanoshells are about 1000 nm. So, it is very unlikely that nanoparticles could accumulate to 'clog' a capillary. As a final indication of size differences, calculations reveal that the volume of a typical protein is between one-millionth and one-billionth of the volume of a typical cell.

Question 3.9

(a) This description refers to nanoshells, as developed by Naomi Halas. They are 'tunable' in that they can be constructed in ways that make them reactive to infrared light. Their activation results in local heating of cancerous tissue, which can eliminate tumours. The term 'precision-guided' reflects the fact that the addition of specific antibodies to their surface means that nanoshells can 'home in' on particular target areas.

(b) It is certainly true that these nanoparticles are non-toxic compared with those chemicals used in conventional chemotherapy for cancer treatment. This is certainly an advantage; the drugs used in chemotherapy kill not only the target cancer cells, but also a significant number of healthy cells, i.e. they are cytotoxic; see Section 7.4.2 of Topic 6. There is no evidence that nanoparticles do so; the drugs they may deliver act on healthy cells too, but their mode of delivery is such that they are delivered 'on target'. Despite this promise, it is too early to claim that such nanoparticles have no side effects.

Question 3.10

The blood–brain barrier is a consequence of the restricted permeability of the cells that form the inner lining of the capillary vessels that permeate the brain. The unusual ability of nanoparticles to cross this barrier is of practical benefit, in that associated drugs might be carried into brain tissue. But the same property raises a more general safety fear, which is that nanoparticles injected into the body (nanoshells, for example) might be conveyed to the brain in the blood and exert unforeseen effects on brain tissue.

Question 4.1

Science communication. The comments on this subject were explicit. Scientists do not always know how to communicate with the public, and scientific jargon is off-putting to a lay audience. It is not clear whether the disagreements among the scientists that the jury noticed were due to lack of communication (e.g. between scientists working in very different fields, though all under the nanotechnology banner) or whether they were the result of scientific differences of opinion (e.g. different interpretations of the same sets of data, different estimates of the time-scale for particular applications to come to the market, etc.).

Risk. The jury made no specific mention of risk, although the recommendation about labelling probably relates to concerns about possible hazards associated with nanoparticles. However, the generally positive attitude 'provided safety could be guaranteed' begs a much larger question. Safety can of course never be totally guaranteed, so one could argue that the jury did not really get to grips with the risk–benefit balance. The comments that even scientists did not always agree and were not sure about the future direction of the field might be taken as an implying that unforeseen and hazardous consequences could result.

Ethical issues. The issue of free access for all to nano-enabled drugs is an ethical one. Presumably the jury's concern was that such drugs might be expensive and, therefore, might be made available only on a selective basis. Ultimately, of course, the funding and operation of the NHS is a matter of politics.

Decision making. The issues of research funding, education and the regulation of labelling on consumer products all involve decision making at governmental level. The recommendation that there should be clear labelling in plain English on products containing manufactured nanoparticles presumably relates to the desire for consumers to be able to exercise informed personal decision making in whether or not they buy such products, and not to buy them unwittingly.

Question 4.2

Points you might have made include the following, though this list is not exhaustive:

- The NanoJury and the *GM Nation?* debates both had similar aims.

- By the time the *GM Nation?* debate took place, GM products were already close to commercial use in the UK and many members of the public had quite firm views about GM technology. The NanoJury, on the other hand, was held at a time when the general public had rather low awareness of nanotechnologies.

- Most of the participants in the *GM Nation?* meetings were self-selected; this may have meant that they came to the debate with views already formed, rather than approaching it impartially. The members of the NanoJury were more representative of the general public, and more ignorant of the technologies at the start of the exercise; however, they were all drawn from one geographical community.

- Most participants in the *GM Nation?* debate attended one meeting each, lasting a few hours. The NanoJury process took a lot more time, with participants spending 25 hours on preparatory activities and 25 hours on the discussion of nanoscience and nanotechnologies. The sessions were spread over many weeks, so allowing time for reflection between meetings.

- The *GM Nation?* debate was set up in a rather confrontational way, with the stimulus material being identified as either 'for' or 'against' GM. There was no particular attempt to inform participants about the underlying science. The feedback form supplied to participants involved a specific list of questions and was drawn up by 'experts'. In the NanoJury experiment, the jury members exercised a large measure of control over the process and the outcomes. They could ask any questions they wanted of the scientific witnesses and draw up their report in their own words.

- Participants in the *GM Nation?* debate were sceptical as to whether their views would have any effect on commercial interests and decision makers, despite the scale of the exercise, because the technology and products were already well established by the time the debate took place. The NanoJury exercise was carried out when the technologies were only just beginning to emerge, but there was no formal mechanism for its deliberations to be fed into decision making processes, merely a promise that its report would be considered by an appropriate UK governmental committee. It is possible that the very small scale of the exercise may militate against it being influential with decision makers.

- The NanoJury UK certainly fulfilled the call made in the Royal Society and Royal Academy of Engineering report for a 'constructive and proactive debate', and was held sufficiently early for there to be few 'deeply entrenched or polarised positions' among the general public (although the scientific witnesses did hold a spectrum of specific views). In principle, the NanoJury could have been in a position to 'inform key decisions', although at the time of writing (early 2006) there is no evidence that this will in fact be the case. The small scale of the exercise may result in it being largely ignored by decision makers.

Comments on activities

Activity 1.1

(i) The total number S of surface atoms is

$$S = 2(n + 1)^2 + 4n(n - 1)$$
$$= 2(n + 1)(n + 1) + 4n(n - 1)$$
$$= 2(n^2 + 2n + 1) + 4n^2 - 4n$$
$$= 2n^2 + 4n + 2 + 4n^2 - 4n$$
$$= 6n^2 + 2$$

(ii) The total number I of interior atoms is

$$I = (n - 1)(n - 1)^2 + n^3$$
$$= (n - 1)[(n - 1)(n - 1)] + n^3$$
$$= (n - 1)[n^2 - 2n + 1] + n^3$$
$$= (n^3 - 2n^2 + n - n^2 + 2n - 1) + n^3$$
$$= 2n^3 - 3n^2 + 3n - 1$$

(iii) For $n = 3$:

$$S = (6 \times 3^2) + 2 = (6 \times 9) + 2 = 54 + 2 = 56$$
$$I = (2 \times 3^3) - (3 \times 3^2) + (3 \times 3) - 1$$
$$= (2 \times 27) - (3 \times 9) + (3 \times 3) - 1$$
$$= 54 - 27 + 9 - 1$$
$$= 35$$

These values for S and I agree with those in Table 1.1, as required.

(iv) The completed version of Table 1.1 is shown in Table 1.4.

Table 1.4 Proportions of surface and interior atoms for a cube made up of $n \times n \times n$ BCC-type blocks

n	Number of blocks in cube ($n \times n \times n$)	Number of surface atoms	Number of interior atoms	Total number of atoms	Proportion of atoms on the surface
1	1	8	1	9	89%
2	8	26	9	35	74%
3	27	56	35	91	61%
4	64	98	91	189	52%
5	125	152	189	341	45%
6	216	218	341	559	39%
33	3.6×10^4	6 536	68 705	75 241	8.7%
100	1.0×10^6	60 002	1 970 299	2 030 301	3.0%
1000	1.0×10^9	6 000 002	1 997 002 999	2 003 003 001	0.3%

(v) A $4 \times 4 \times 4$ cube is the largest one in which more than half the atoms (52%) are at the surface. For iron, this cube would have an edge length of 4×0.3 nm = 1.2 nm.

For iron, the cube with an edge length nearest to 10 nm has $n = 33$ (i.e. edge length 33×0.3 nm = 9.9 nm). Roughly 9% of atoms are on the surface. The largest cube covered by the data in the table has $n = 1000$ (i.e. edge length 1000×0.3 nm = 300 nm) and has only 0.3% of its atoms on the surface. It is debatable whether even this can be described as a bulk solid – it is certainly very tiny! Extrapolating from the trend in Table 1.4, it is clear that for a true bulk solid the proportion of atoms on the surface is likely to be only a small fraction of a per cent.

Activity 2.1

Communication. All of Schön's work got past peer review. Whilst the editors of *Science* and *Nature* did not feel that the case exposed flaws in the peer-review system, you might not agree. Should peer review aim to detect fraud? Should the specialist editors at the journals be looking for fraud or inconsistencies? What about communication between scientists? Why did co-authors not see the experiments performed to which their names were attached? Were reviewers and co-workers nervous of being too critical of such stunning results from such a prolific and highly rated researcher? For, although peer review is intended to be an anonymous process, in small and specialised fields this can be hard to achieve.

Ethical issues. There are ethical issues simply around the fact that Schön faked his results. You might also have thought about the pressure that young researchers feel under to make progress in a field where there is commercial interest. A good deal of money clearly followed Schön once he began publishing such extraordinary results. There are issues around the amount of freedom Schön was given, and how little scrutiny his work received from his co-authors. Towards the end of the article the author suggests that biologists have a better system for preventing and investigating scientific misconduct. Whilst researchers working with living subjects do need to have their research plans approved by ethics committees, these are concerned with different issues than scientific fraud, and there are cases of scientific misconduct in the biological sciences as elsewhere. There are ethical issues involved in considering any system that attempts to prevent such misconduct.

Activity 2.2

Your answer to this question may depend on what year you are studying this course. You are not expected to know the answers but at the time of writing (2006) all are 'on the table' as potential or actual developments. Nanotube fuel cells powering electronics and vehicles, nano-engineered cochlear implants and intelligent clothing are still at the 'wishful' stage. Organic LEDs, photovoltaic films, coated fabrics and hip joints made from biocompatible materials are at an early stage of development. All the others, whilst maybe still being developed further, are well advanced.

Activity 2.3

(a) Vector A is horizontal across the sheet. If n and m were equal, then the tube will always have an armchair configuration. The smaller the value of n and m, the smaller the diameter of the tube (Figure 2.25).

(b) Vector Z goes up the page at 30° to vector A.

(c) You can make zigzag tubes by keeping either *m* or *n* equal to zero.

(d) You should have found that all chiral nanotubes have chiral angles somewhere between 0° and 30°.

Figure 2.25 Construction of armchair and zigzag chiral vectors.

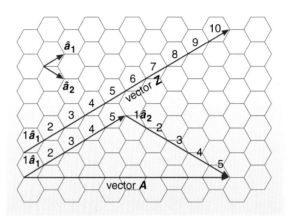

Activity 3.1

(i) Because they are at this stage single stranded, i.e. they have exposed bases. They are identical to what was called sticky ends in Topic 6, Section 3.1.

(ii) Given that the robot's sticky feet are single stranded, as are the spikes or footholds of the walkway, then the anchor strands must also be single stranded, since they bind the footholds and the feet together. When the base sequences of particular feet and footholds are complementary to particular anchors, binding will ensue. The only way in which sticky feet will bind to *particular* footholds is if each binding is specific, so each anchor strand must have a unique base sequence, which explains why the anchors are differently coloured in (b).

(iii) The yellow strand must have a base-pair sequence that is complementary to that of the right anchor. The reason binding is sufficient to strip off the anchor from its existing binding is similar to that described in relation to the molecular tweezers in Figure 3.7: the greater the degree of pairing possible, the greater the likelihood of binding between two strands. (Figure 3.8 reveals that each anchor has an upper handle which does not bind to the foot or foothold. The yellow strand sticks to this handle and then to all the other bases of the anchor.)

(iv) For the free foot to become bound onto foothold C, it needs to become attached to another, specific anchor strand, shown in green in Figure 3.8d. The base sequence of this anchor strand needs to be complementary to parts of the leading robot leg and to foothold C.

(v) Another competing strand (shown in light blue in Figure 3.8e) has to strip the left leg from foothold A. The free leg will bind to foothold B if a new anchor strand (shown in grey in Figure 3.8f) becomes positioned as shown. At this stage the entire robot has taken one step forward!

Now read through Figure 3.8 to ensure you follow the procedure. Incidentally, you might ask how the researchers know these events occurred, given (as with Figure 3.7) this occurs on a scale too small to be visualised. The various DNA anchors and stripping strands were added one by one, and samples of the solutions containing the nanorobots were taken. These samples were then subject to electrophoresis (Section 3.1 of Topic 6), which could identify double-stranded DNA of the type expected if the events of Figure 3.8 took place.

Activity 3.2

I would not support the idea of the research being suspended, given the potential gains from the safe application of the technique, but I am persuaded by Mae-Wan Ho's words about our our lack of knowledge about safety issues. For example, might quantum dots prove toxic if they were retained in the body? You'll appreciate from Chapter 1 that, after inhalation, nanoparticles can become distributed widely in the body, not just in the lungs. It is striking that nanoparticles are non-biodegradable, which does open up the possibility of their transport in the blood and eventual accumulation after injection, perhaps in the liver.

Activity 3.3

1   A great range of potentially valuable applications of nanotechnology are emerging in medicine, e.g. in diagnostics and drug delivery.

2   With high development costs and stringent safety assessment, many of the hoped-for products of nanotechnology are unlikely to come to fruition; most techniques in development have strong advocates, sometimes prone to overstatement.

3   Self-assembling nanotubes consisting of stacked cyclic peptides have the capacity to disrupt the membrane of bacteria such as MRSA. The chemical make-up of the constituent amino acids influences the mode of action and specificity of such nanotubes; there may be a reduced likelihood that bacteria develop resistance to such agents.

4   Immunological responses to implanted insulin-producing pancreatic cells can be avoided by enclosing such cells in silicone capsules with walls permeated by nanoscale pores. As a medium-sized protein, insulin can pass through the pores, following the stimulation of its release by the influx of glucose.

5   Nanoshells and dendrimers are manufactured nanoparticles that have potential for delivery of other agents. Both could be used for locating and treating tumours. With nanoshells, this involves either direct local heating after infrared stimulation and/or heat-provoked release of contained drugs. Complex, multifunction dendrimers have the potential to diagnose, locate and treat tumours; dendrimers tagged with iron oxide can be readily visualised.

6   Buckminsterfullerene, when complexed with a dendrimer, has the capacity to block an enzyme that is central to the replication of the HIV virus.

7   Nanoparticles have the exceptional ability to cross the blood–brain barrier and are of potential value in helping deliver pharmaceuticals directly to brain tissue to alleviate neural diseases.

8   The safety implications of injecting nanoparticles into the body remain largely unexplored and will have to be comprehensively researched, if the claimed benefits of nanoscale treatments are to be achieved safely and reliably.

Activity 3.4

In most developments, you will find evidence of the beginnings of commercial exploitation. For example, Desai works in part for a 'private bio-engineering company'; Naomi Halas refers to excitement 'within the company' of work to date on nanoshells; the firm C-Sixty has been founded to exploit the potential use of buckminsterfullerene. There is also mention of the interest of pharmaceutical companies in nanoscale products that can cross the blood–brain barrier.

References

Alivisatos, P. (2001) Less is more in medicine, *Scientific American*, **285**, pp. 59–65.

Brooks (2000) All aboard the nanotube, *Worldlink*. Available from: http://www.worldlink.co.uk/discuss/msgReader$227 [Accessed April 2005].

Chemical and Engineering News, **81**, pp. 39–42

Drexler, K. E. (1986) *Engines of Creation: the coming era of nanotechnology*, Anchor Books.

Drexler, K. E. (2001) Machine-phase nanotechnology, *Scientific American*, **285**, pp. 74–75.

Editors of Scientific American (2002) *Understanding Nanotechnology*, Scientific American Inc. and Byron Press Visual Publications Inc., Warner Books.

European Commission (2004) *Towards a European Strategy for Nanotechnology*.

Faraday, M. (1857) The Bakerian Lecture. Experimental relations of gold (and other metals) to light, *Philosophical Transactions of the Royal Society of London*, **147**, pp. 145–181.

Goodsell, D. S. (2004) *Bionanotechnology: lessons from nature*, Wiley.

Ho, M-W. (2004) Metal nanoshells; cure or curse? *Science in Society*, issue 21. Available from: http://www.i-sis.org.uk/isisnews/sis21.php [Accessed April 2006].

Health and Safety Executive (2004) *Health effects of particles produced for nanotechnologies*, HSE Hazard Assessment Document, EH 75/6, December 2004.

Jones, R. A. L. (2004) *Soft Machines: nanotechnology and life*, Oxford University Press.

Kelleher, K. (2003) Engineers light up cancer research, *Popular Science*, 6 November.

King, D. (2003) Foreword. In Wood, S., Jones, R. and Geldart, A., *The Social and Economic Challenges of Nanotechnology*, Economic and Social Research Council.

New Scientist (2002), Conduct unbecoming, **2363**, 5 October 2002, p. 3.

Nova Science Now (2005) Working with Nanoshells. A conversation with Naomi Halas. Available from: http://www.pbs.org/wgbh/nova/sciencenow/3209/03-nanoshells.html [Accessed April 2006].

Ratner, M. and Ratner, D. (2003) *Nanotechnology, a gentle introduction to the next big idea*, Prentice Hall.

Royal Society and The Royal Academy of Engineering (2004) *Nanoscience and Nanotechnologies: opportunities and uncertainties*.

Taylor, J. M. (2002) *New Dimensions for Manufacturing: A UK Strategy for Nanotechnology*. Report of the UK Advisory Group on Nanotechnology Applications submitted to Lord Sainsbury, Minister for Science and Innovation, Department of Trade and Industry/Office of Science and Technology.

Acknowledgements

The Topic 7 authors are grateful for the assistance of Yvonne Ashmore in drawing chemical structures and of Mark Hirst in producing molecular models.

Grateful acknowledgement is made to the following sources for permission to reproduce material within this book:

Cover image: Reprint courtesy of International Business Machines Corporation copyright 1993. © International Business Machines Corporation.

Figure 1.3a: © NASA; *Figure 1.8*: Max-Planck-Institut fur Metallforschung; *Figure 1.9*: Crommie, M. F., Lutz, C. P. and Eigler, D. M. (1993) 'Confinement of electrons to quantum corrals on a metal surface', *Science*, **262**, pp. 218–220. Reproduced with permission IBM Research, Almaden Research Centre; *Figure 1.10*: © Courtesy of IBM Zurich Laboratory; *Figure 1.13*: Buffat, P. and Borel, J. (1976) 'Size effect on the melting temperature of gold particles', *Physical Review*, June 1976. © 1976 American Physical Society; *Figure 1.17*: Mark Dadswell/Allsport, by permission of Getty Images; *Figure 1.18a*: Alain Rochefort Assistant Professor, Engineering Physics Department, Nanostructure Group, Center for Research on Computation and its Applications (CERCA); *Figure 1.19*: Donaldson, K. (2004) Royal Society and Royal Academy of Engineering Report , The University of Edinburgh, reproduced with permission; *Figure 1.22*: Adapted from http://www.ninger.com; *Figure 1.23a, b*: Adapted from Freemantle, M. (1999) 'Blossoming of dendrimers', Chemical and Engineering News, **77** (4) © 1999 American Chemical Society; *Figure 1.23c*: Adapted from Freitas Jr, R. A., *Nanomedicine*, Vol 1, 1999. Landes Bioscience; *Figure 2.5*: Cox, J. (2003) 'A quantum paintbox', *Chemistry World*, September 2003. The Royal Society of Chemistry; *Figure 2.6*: Evident Technologies; *Figure 2.7*: Service, R. F. (2002) 'Research integrity: pioneering physics papers under suspicion for data manipulation', *Science*, **296**, p. 5572, 24 May 2002. Copyright © 2002 AAAS; *Figure 2.9*: Poole, C. P. and Owens, F. J. (eds) *Introduction to Nanotechnology: Selected Topics*, 2003. John Wiley & Sons, Inc; *Figure 2.10*: Phase Transformations and Complex Properties Research Group, Department of Materials Science and Metallurgy, University of Cambridge; *Figure 2.20*: Dresselhaus, M. et al. (1998) 'Carbon nanotubes', *Physics World*, January 1998. Institute of Physics Publishing; *Figure 3.2a, b and d*: Mader, S. S. Biology, 2001. McGraw-Hill Companies Inc; *Figure 3.2c*: Stryer, L. (1988) 'Part 1, molecular design for life', *Biochemistry*. 3rd edn, W. H. Freeman and Company; *Figure 3.4a*: Courtesy of Professor Mike Stewart, Department of Biological Sciences, The Open University; *Figure 3.4c*: Alberts, B. and Bray, D., *Molecular Biology of The Cell*, 4th edn 2002, Garland Publishing, Inc.; *Figure 3.5 and 3.10*: Goodsell, D. S., *Bionanotechnology*, 2004. Wiley–Liss Inc. A John Wiley & Sons, Inc. Publication; *Figure 3.7*: Copyright © 2006 Lucent Technologies; *Figure 3.8*: Hogan, J. (2004) 'DNA robot takes its first steps', *New Scientist*, 6 May 2004; *Figure 3.12*: Professor Naomi Halas; *Figure 3.13*: Michigan Center for Biologic Nanotechnology.

Table 3.1: Poole, C. P. and Owens, F. J. (eds) *Introduction to Nanotechnology*, 2003. John Wiley & Sons, Inc.

Activity 2.1: 'Conduct unbecoming' (2002) *New Scientist*, Issue 2363, 5 October 2002.

Every effort has been made to contact copyright holders. If any have been inadvertently overlooked, the publishers will be pleased to make the necessary arrangements at the first opportunity.

Index

Entries in **bold** are key terms defined, along with other important terms, in the Glossary. Page numbers referring only to figures and tables are printed in *italics*.